FROM IDEA TO PROFIT
Managing Advanced Manufacturing Technology.

FROM IDEA TO PROFIT
Managing Advanced Manufacturing Technology.

Jule A. Miller

VNR VAN NOSTRAND REINHOLD COMPANY
New York

Library of Congress Catalog Card Number: 85-17975
ISBN: 0-442-26083-0

Manufactured in the United States of America

Published by Van Nostrand Reinhold Company Inc.
115 Fifth Avenue
New York, New York 10003

Van Nostrand Reinhold Company Limited
Molly Millars Lane
Wokingham, Berkshire RG11 2PY, England

Van Nostrand Reinhold
480 Latrobe Street
Melbourne, Victoria 3000, Australia

Macmillan of Canada
Division of Gage Publishing Limited
164 Commander Boulevard
Agincourt, Ontario M1S 3C7, Canada

15 14 13 12 11 10 9 8 7 6 5 4 3 2 1

Library of Congress Cataloging-in-Publication Data

Miller, Jule A., 1935–
From idea to profit.

Bibliography: p.
Includes index.
1. Manufacturing processes—Technological innovations—Management. I. Title.
TS183.M55 1986 658.5′14 85-17975
ISBN 0-442-26083-0

This book is dedicated to my father, to my son, to Peter Zumsande, and to all the members of the Secret Society of Those Who Make Things Work.

"I have but one lamp by which my feet are guided,
and that is the lamp of experience."
Patrick Henry, 1775

PREFACE

In the spring of 1982 I attended a conference on laser innovations in manufacturing sponsored by the Society of Manufacturing Engineers. Several of the first morning's presentations were given by equipment manufacturers. These followed the usual format of such talks; the first slide shows a particular product or component, and the speaker briefly explains its function and the problems in its manufacture. The second slide shows the new machine or production line which solved all the problems, reduced costs, improved quality, and, the speaker assures us, is now operating flawlessly in his customer's shop. The third slide shows another product, and the next shows the facility built to make it, and so on.

Over lunch we were discussing the more clever of these applications, when someone said that elegant ideas are relatively easy to find; the real problem is to get from the odd-numbered to the even-numbered slides. This step, between conceiving a manufacturing technology improvement and implementing it in production, is the key to a company, or a country for that matter, gaining and maintaining a competitive advantage. It is very poorly understood. The engineering staff sees such projects as a series of technical problems to be solved, and sometimes considers management and obstruction to be synonymous. Corporate, divisional, and financial managers tend to see these programs as a series of milestones and budget objectives to be met, often without understanding the possibly critical technical implications of what appear to be straightforward business decisions or policies.

For these projects to succeed, that is, for the improved manufacturing technique to be implemented in the shortest time, at the least cost, and then to produce the most profit, the often contradictory requirements of technical reality and standard management procedures must be reconciled. Far more of these projects fail because of administrative blunders than because of technical errors. For example, starting from a potentially excellent application of a given technology, one could conduct a rational feasibility study and, when it succeeds, select a contractor to build the production process equipment. Many successful manufacturing technology programs and many manufacturing technology disasters have been conducted by this method. Other approaches are possible, such as building a pilot plant in one's own laboratory before subcontracting the final system.

Obviously, not all programs are alike. There are important, and sometimes subtle, technical differences between manufacturing processes which make a program management technique work in one case, yet fail in another. One can be sure, however, that when the failure occurs, the technology is always blamed, no matter how poorly the program was conducted or how unsuited to the particular problem was the program management technique selected.

That is tragic. Not only is a manufacturing technology which might improve a company's productivity and competitive position lost, but a technical approach that might be of great value in other areas of the company's operations is discarded and sometimes permanently blackballed.

It is the purpose of this book to improve the efficiency of these projects and to reduce the chances of costly failures by providing those in charge of manufacturing technology projects with practical information on their organization and conduct. Although the standard planning and project control techniques such as PERT and CPM are discussed briefly, these are not the focus of this volume, nor are they what is meant by managing advanced manufacturing technology. Rather, this book is concerned with the specific steps that must be considered in a properly run feasibility study, development program, and introduction into production, and with how these things are affected by the realities of the modern American company.

It is primarily written for the practitioner, whether he is called the project manager, supervisor, principal investigator, or resident engineer. It is assumed that he is a technical person managing a technology-based effort. If he does not completely understand the technical basis of the program, neither this nor any book on management will help him or the project very much.

Other levels of management must also understand the problems and options facing the project manager. Without this understanding, they will have great difficulty in selecting projects, allocating resources, assessing risk, and deciding if a program is succeeding. In some cases it will be difficult to know if the project was successful even after it is completed. It is surprising how many supposedly successfully implemented innovations can be found two or three years later, unused, in some back corner of the shop. Investigation usually shows that once all of the show and tell was over, and top management's attention had gone to other things, the manufacturing department quietly discarded the new technique because it was not truly production suitable or not properly implemented by those who developed it.

Upper management, in many cases, does not have the specialized technical background to independently evaluate progress in manufacturing technology development. If, in addition, management does not understand how these programs should be planned and conducted, it is truly unable to perform its function of overall control. One who understands neither what is being done nor how it should be done has small chance of influencing or even understanding the outcome.

The advancement of manufacturing technology is a very risky business. There are three basic ways it can gravely injure a company:

1. Resources can be wasted on inappropriate, technically hopeless, or badly managed projects.
2. Resources can be wasted on programs that develop innovations which appear to succeed but which produce a product of such poor quality that it injures or even destroys the company's competitive position.
3. One can avoid the first two risks by sticking to established technology until an opportunity to advance without risk comes along. In the meantime, management can issue general policy statements about dedication to innovation, while disapproving every specific project due to excessive risk. This policy can continue until the announcement of the closing of the plant because it can no longer compete.

Management is the reconciling of conflicts and minimizing of risk. However, conflict, risk, and uncertainty can be eliminated only by liquidating the business. This book is intended to provide an understanding of the conflicting factors in advanced manufacturing technology development, so they can be logically reconciled.

The exact definition of advanced manufacturing technology will vary from industry to industry and company to company. One company's radical innovation will be another's mature technology. Obviously, the development of an advanced technology and the procurement of additional facilities for a mature one can and should be managed by different techniques. The technology this book addresses is assumed to be advanced for the particular company, to be a significant technological step forward in the way it manufactures its products. It is also assumed to be a step which is commenced with an imperfect understanding of all that is involved in reaching the goal of an implemented manufacturing process. If everything is understood at the beginning, it is mature, not advanced, technology.

The environment in which the development and implementation of advanced manufacturing technology takes place is the modern American corporation. It is not possible to write a truly useful book on this subject without dealing with the realities of that environment. There are some excellent companies with corporate policies and cultures which facilitate innovation and implementation of improved manufacturing technology. At the other end of the bell curve are companies which, while often chanting slogans about progress, have corporate systems which place so many control points, approvals, justifications, milestones, reports, and committee reviews in the way that no idea and no project manager can possibly succeed.

The environment this book generally assumes is the company in the center of the distribution: reasonably well run, and not yet destroyed by the attack of the killer bean counters. In such companies the road to innovation is neither impossible nor easy. It is difficult enough to make a truly successful program one of the most personally rewarding experiences found anywhere in professional life.

J.A.M

ACKNOWLEDGMENTS

The author wishes to thank the management and employees of the Avco Lycoming Division for the support given, not only in the preparation of this book, but in all of the projects, over the years, which led to it.

In particular, I would like to thank Jeremy Walters and Louis Fiedler for their help and support and acknowledge the contributions of the entire Materials Laboratory staff to this book and the advancement of manufacturing technology it reflects.

The special contributions of Gerard Bessette, John McMurray, Mary Bazan, Jack Lee, and Ken Struzek are gratefully acknowledged, as are those of Nancy Moran and Millicent Hanson in the preparation of the original proposal and typing of the final manuscript, and Anthony Gravino and Joseph Gallagher for photographic services.

Many other people and companies contributed to the projects described in this book and the author wishes to thank all of them. In particular, the contributions of Don Bowes and the staff at Coherent, Incorporated are gratefully acknowledged.

Finally, I wish to thank Tim Foley for his tolerant ear and encouragement, and my wife, Heide, who, as in everything, was a full partner in the writing of this book.

J.A.M.

CONTENTS

1 INTRODUCTION TO THE MANAGEMENT OF ADVANCED MANUFACTURING TECHNOLOGY

ORGANIZATIONS

The management of advanced manufacturing technology is the management of radical change in organizations which place great value on continuity. Most manufacturing companies seek to maximize profit by long production runs, smooth flow of parts and assemblies, steady cost reduction and cost avoidance. Man-tech development is disruptive and, in the short run, expensive. Although its short-term cost and its inherent riskiness have been blamed for the failure of many companies to give the development and adoption of new manufacturing techniques more than just lip service, the fear of disruption is at least as important.

Its manufacturing technology is the basis of a manufacturing company. The way in which functions are divided between departments and how those functions are performed result from what the company makes and how it makes it. Who succeeds and who fails, who has real power and who has only apparent authority are also eventually traceable, no matter how circuitously, to the realities of the production floor and the demands those realities place upon the organization.

Therefore the organization of the company is basic to a discussion of the management of advanced manufacturing technology. Because the details of departmental organization vary from company to company, a generic structure is given in Figure 1.1. *Company,* for the purposes of this book, refers to a complete manufacturing business. It might be a separate corporation or a division or subdivision of a larger entity.

The departments listed in the figure will exist in practically any manufacturing business except the very smallest, and even there, all of the functions of the departments will be done, even if all by the same person.

Within each of these major departments will be various levels of management, ranging from the senior departmental manager, who may be a vice president, down through middle management to supervision. The number of these levels will vary with the size and, to a large extent, the

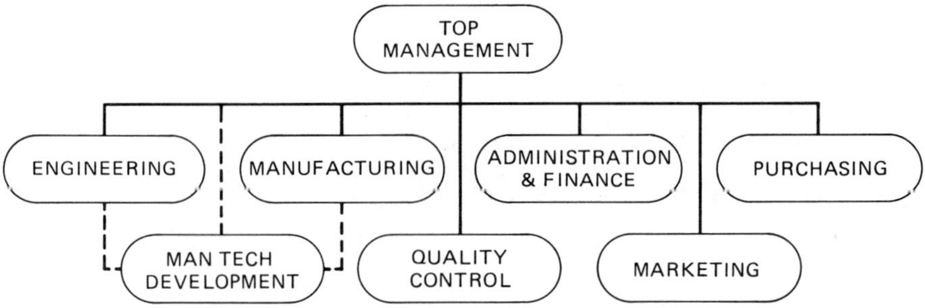

Figure 1.1. Generic Departmental Structure of a Manufacturing Company.

history and culture of the particular company. For the purposes of this book, however, this description of the horizontal division of major tasks and the vertical division of authority will suffice.

The advanced manufacturing technology development unit may be part of a manufacturing engineering function and thus report ultimately to manufacturing operations, or it may be a part of the engineering department. In some rare cases, it might report to general management as a separate independent operation. However, the tasks and problems of a man-tech department and the manager who heads it will be essentially the same, no matter from which branch of the organization they are hung. This group has the central position in improving a company's manufacturing technology, and this book is primarily concerned with how it should do its job.

The man-tech manager and project manager operate across interfaces. It is their job to select basic technology, usually invented by someone else, develop a manufacturing process from it, and by some poorly defined process, implement it.

Implementation is the goal of man-tech programs. It is completed when a production system is operating on the shop floor, producing an acceptable product, and meeting a production schedule while operated and maintained by production labor, working under production supervision, earning a profit.

TYPES OF MANUFACTURING TECHNOLOGY DEVELOPMENT PROGRAMS

Man-tech development projects can be classified as follows:

1. A new way of performing existing operations on an existing part, for example, replacing a manual lathe with a computer numerical controlled (CNC) one.

2. A new operation replacing an old one in an existing process for an existing part, for example, replacing resistance welding with laser welding.
3. An entirely new way of making an existing part, for example, introducing an automated flexible machining line.
4. An improved part made by a new process which could not be made except by the new process, for example, replacing a mechanically assembled turbine engine compressor with a one-piece, electron beam welded assembly.
5. Modifications and improvements to software and computer networks which facilitate the control of manufacturing processes; for example, computer aided design–computer aided manufacturing (CAD-CAM) systems which eliminate the need for drawings and process sheets by allowing the design computer to generate the programs for the machine tool computers.

Each of these types of programs presents different management problems as well as technical problems. Each may require different development program schemes in order to be successful. Some overall outline of the development process must be defined, however, if only to give a framework upon which to construct a system for dealing with the questions in a rational order.

THE PROCESS OF ADVANCED MANUFACTURING TECHNOLOGY DEVELOPMENT

The obvious sequence of operations in the development process is shown in Figure 1.2. The first box contains the discovery of some scientific principle, such as laser theory and its laboratory demonstration. Hopefully, it will also contain the development of devices suitable for production applications. This is the research part of the process and is not usually considered part of the man-tech development process. However, information acquired in the later steps by those trying to use the device for industrial applications should be fed back into the research block where it can influence the path of future work.

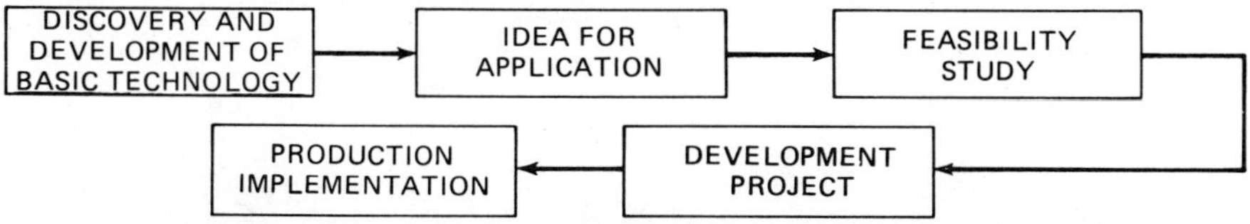

Figure 1.2. The Manufacturing Technology Development Process.

The second box is normally considered the beginning of the advanced manufacturing technology development cycle. Someone who knows about the development of a new device suggests that it might provide an improved manufacturing technique or solve a design problem. Some investigation, often involving a simple experiment, is conducted. Then the actual steps in the development process begin.

In the third box, an inexpensive feasibility study is designed and conducted to ascertain if the process actually is applicable to the case being considered. If the results of this preliminary study are favorable, a program to develop the process is conducted (box 4). If the development is successfully completed, the improvement is introduced into production (box 5), and manufacturing shop personnel are trained in its use and eventually accept it from its developers.

This is, of course, a simplified skeleton of the process, and it does not even contain all of the bones, much less the flesh and blood. In order for someone to find that lasers exist and make the supposition that they might offer a solution to a problem, he has to know that the problem exists. This means he has some understanding of the existing product and the manufacturing process for making it. This is the user pull theory of innovation. There are also new technologies which are so obviously better than anything previously available, that uses for them are actively sought. This is the technology push theory. In reality, both probably meet in one mind when an idea for a manufacturing technology improvement occurs (Figure 1.3).

Figure 1.3 is a better model because it shows that for an idea to occur, both insight into the product and its manufacture and awareness of the new technology are needed. Very few projects implement the original idea as conceived. It is constantly modified and improved as the project continues. This is, after all, why projects are conducted.

Figure 1.4 is thus a still better model because it admits that ideas do

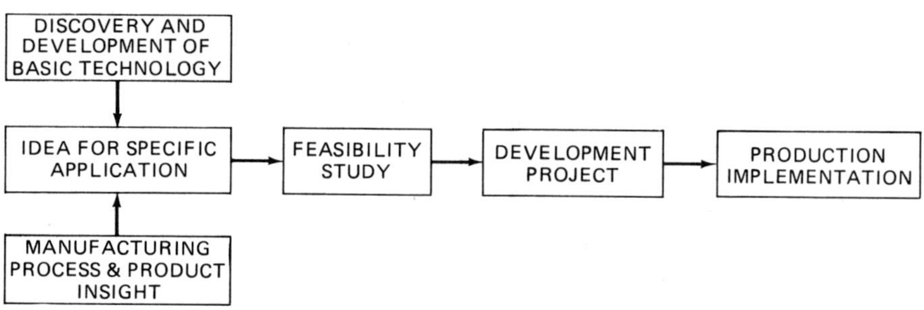

Figure 1.3. The Manufacturing Technology Development Process.

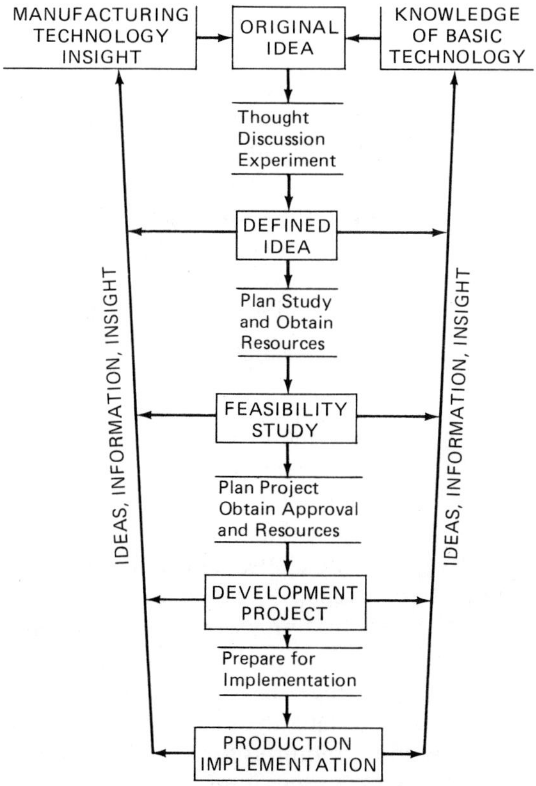

Figure 1.4. The Manufacturing Technology Development Process.

not happen in neat little boxes. It shows that what happens between the boxes is generally more difficult, in terms of management, than what is done in the boxes. At the arrows, decisions whether to continue or give up the project must be made. If it is to continue, it is between the boxes that one is seeking funding and approvals, deciding what resources to commit, and attempting to plan the next major phase so that the most progress can be gained in the least time, for the least cost, and at the least risk.

It is, of course, impossible to list in the arrows all of the between-phase functions. This diagram also attempts to illustrate the constant transfer of ideas, information, and process insight back and forth. Process insight and information are not the same. Information is organized data that are readily transferred and understood. Such things as machine tool feeds and speeds, heat-treating times and temperatures, and material specifications are examples. Process insight is a great deal more difficult to define, to acquire, and to transfer. It is an understanding of the subtle interrelationships within a process that are often not readily obvious or quantifiable. It is the ability of a machine operator to know that something is about to go wrong with a process before there are any tangible signs of trouble. It is the ability of a good welding technician, when faced with a new part of a material and configuration that he has not seen before, to do some crude calculations, look at the part, stand jingling the change in his pocket for a few seconds, and then tell the welder to try a certain combination of parameters which produce a perfect weld.

Process insight is usually impossible to quantify and is, therefore, often ignored in discussions of business administration. It is critically important to all stages of man-tech development and is a major product of all successful projects.

The key to finding good manufacturing technology ideas is to find lots of ideas, try them all, then pick the best. A well run man-tech development department will do many quick, clever experiments aimed at finding if implementation of an idea is possible, fewer feasibility studies, and still fewer development projects. The chances of failure to implement will also drop steeply as projects progress.

It is important that the very first phase of a project include some simple, objective experiment to provide a preliminary evaluation of the idea as soon as possible: for example, making six inches of weld by a new process to assure that it is capable of joining the materials in question. No technical or cost justification should be required for this experiment, and the man-tech manager should have a standing fund for this, needing no approvals.

In a tightly run, rigidly controlled company, it is impossible to justify spending any time or money on something never before tried. In order to write a justification one has to know, or at least claim to know, something. A new idea is never more than a guess. Just the fact that someone can get a few days and a few dollars to try something, and if it fails not have it reflect on his credibility, is essential to the generation of ideas.

The worst way to evaluate new ideas prior to trying them is by committee. The mere existence of such a group, before which someone with an idea must appear, is enough to chill all but the most aggressive innovators. Such groups usually do not evaluate ideas objectively but base decisions on political factors such as whose idea it was and where he fits into the corporate structure. If an idea is of obvious value, committee members usually will not approve it as suggested but will immediately modify it out of existence by staging a brilliant suggestions contest among themselves. At every step of the development process, the value of an innovation should be evaluated only upon experimental evidence.

When the preliminary experiment has provided some facts about what might be needed to go from the idea to implementation, some justification for the feasibility study should be required. But not too much. Once the feasibility study is completed, and the information and insight needed to define the development and implementation phases are available, the requirements for justification should be much more stringent. But throughout the program, the amount of justification required at each stage should be proportional to the amount of money already spent as well as the amount to be spent.

Once it is decided that an idea has potential, a feasibility study is defined. The first step in this definition is to write, in one sentence, the overall purpose of the program. This sentence, in most cases should start with the words, "The purpose of this program is to implement in production a . . . ." The next step is to similarly define the purpose of the feasibility phase. Then the criteria for success of the study should be stated, and finally the individual tasks are listed with their criteria for success.

A feasibility study is a controlled experiment to assure that the technology being considered for an application has the basic capability of performing the necessary manufacturing operation or operations. It is neither a full-scale process development project nor a quick, one-sample demonstration. Although the preliminary experiment only has to show that the technique being considered can process the material in question, the feasibility study must prove that it is possible for the technique to make an acceptable product. This study should also give at least a preliminary definition of the technical and management approach appropriate to the development phase, which must be managed so it is focused and logically sound without being rigid. Flexibility and open-mindedness are particularly important but should not be mistaken for evasion of the basic questions the study is being conducted to answer.

If the results of the feasibility study justify continuing with the development of an advanced manufacturing technique, the next step is to plan for the development program. Again, the objectives of this phase and the minimum requirements for success should be written, and then the individual tasks defined in the order in which they must be completed. If the feasibility study was properly conducted, the central technical questions that must be answered by the development program should now be known. These must be attacked first. If the proposed system consists of six wheels, a length of rope, and a sky hook, no resources should be put into building the wheels or testing rope samples until it has been demonstrated that the sky hook works dependably. This is not nearly as obvious in a real situation as it appears. The organizational pressures often push one toward showing early progress, and there are always tasks in any development program that require no development, thus providing an easy way to impress those who are more concerned with milestone charts than with true progress.

Once the development phase is planned, it is possible to make the estimates needed to obtain funding. These estimates also make it possible to decide whether something shown to be feasible is also economically desirable. It is possible to conduct an elegant program that takes so long and costs so much that, even if completely successful, it can never yield a profit. No matter what secret compulsions drive one to conduct these programs, increased productivity, quality, and profit are the stated reasons.

The development program is now conducted. This could involve the building and testing of an entire plant or production line, or it could replace just one operation with a more advanced technique.

During the development program, planning is begun for what is usually the most difficult part of the management of advanced manufacturing technology, production implementation. In this phase, the manufacturing operations department is taught to run the new process, service and maintain the equipment, supply it with proper input materials, and inspect its output.

More precise and more complicated models of the man-tech development process can be constructed, but the purpose of this book is not to model or to define, it is to explain, and therefore Figure 1.4 is an adequate outline upon which to organize it. Chapter 2 will discuss the structures of companies and how they affect the innovative process as well as the structures of the departments which conduct these projects and their relationship to the other parts of the organization. Thus it deals with the environment in which the innovative process takes place: the force field of ideas, interrelated policies, insights, data, and politics which comprise the company.

Chapters 3 through 6 discuss the management of the manufacturing technology development program from feasibility study through the final testing of the production suitable equipment. Chapter 7 deals with the problems of implementation.

THE TURN-KEY APPROACH[1]

There is another, totally different way of managing the procurement of new manufacturing technology and introducing it into production: the turnkey system (Figure 1.5). This approach sees the acquisition and introduction of advanced manufacturing technology as a purchasing function. One simply buys it. It has a certain attractive simplicity about it, because

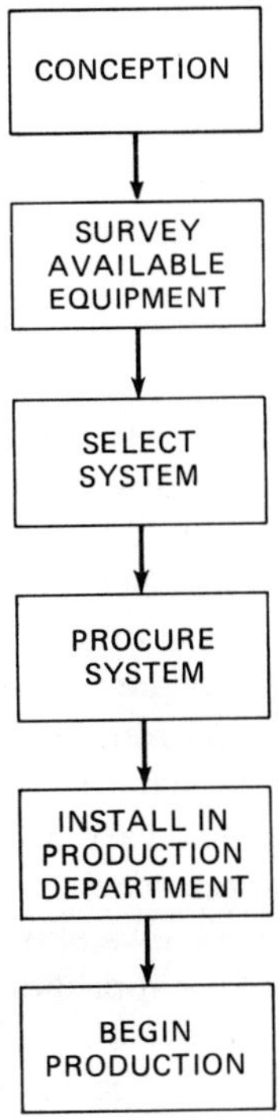

Figure 1.5. The Turn Key Approach to Advanced Manufacturing Technology.

it does not admit to the existence of any of the problems that commonly occur. For very new technology—particularly systems involving computers and robotic functions—with which the company's manufacturing departments are not familiar, the use of the turnkey approach almost guarantees massive implementation problems, if not failure.

This method is how most purchases in the economy are made, and it therefore readily follows the standard corporate paper path. It is decided that some new machine is an improvement over the company's present way of doing something, or an advanced technology system is being introduced by other companies and top management whimsically decides that the company should also have one.

The procurement department is told to buy the necessary equipment. The person assigned to the job, particularly in large companies, often has little or no understanding of the product the machine will make or the overall manufacturing process into which it must fit. Following the normal procurement cycle, he issues a request for quotation to a number of potential suppliers, and when the bids are received, he selects a supplier, usually overemphasizing price because that is the only readily comparable quantity, and places an order. When the equipment is received, it is set up on the shop floor and in theory production begins. The procuring department announces the successful completion of its assignment: The new whatsis, obtained in the fastest, cheapest way, is in place. They then leave the project. The manufacturing departments then must solve all of the questions and problems that were not even considered by the turnkey procurement approach.

This approach is inappropriate to the management of advanced manufacturing technology projects because its basic assumptions are not met in such situations. These assumptions are that both the buyer and the seller completely understand the device being procured and the intended application, and all sellers offer essentially equivalent products. It is appropriate for the purchase of shoelaces, where from the various lengths, colors, and materials available the required pair is selected, factoring in price. If the buyer does not know what length he needs for the particular shoes in question, even in this simple case he may find that he has problems when he attempts implementation. The less the buyer knows about the equipment, and the less the seller knows about its intended use, the greater the unmanaged risk of the turnkey approach.

The second assumption, that all sellers offer essentially equivalent products, is practically never met in the case of advanced manufacturing equipment. Yet, everything else must be equal if final selection is to be made primarily on price. A production machine that cost ten dollars and does not work is infinitely more expensive than one that cost fifteen dollars and does work. If turnkey procurement is used, one does not know what one has bought until it is too late.

There is often an excessive trust in manufacturers' warranties in these projects. The assurance is given by the salesman to the purchaser, in all honesty, that the manufacturer will stand behind the system and make it work on the customer's production floor. If, as is sometimes the case, the equipment is found to be completely unsuitable for the application, no amount of warranty effort will make it work. Having the manufacturer's servicemen and engineers fight the system in the customer's shop for months, often with the customer's maintenance people working with them, will, however, completely alienate manufacturing management. They will feel, with justification, that they have been given not an easier way to meet their production schedule, but an impediment. There are many examples of how this has led to destructive adversary relationships between equipment manufacturers and their customers.

One of the most difficult problems to overcome in the introduction of even a properly developed and proven manufacturing technique is the not-invented-here syndrome of some operations people. This is the conviction that the new system was obtained without their participation or consent by someone who neither understood nor cared about their problems. This problem is aggravated in turnkey procurement. Tools or modifications required by the particular product and methods of the company are often not supplied because the equipment manufacturer is not aware of them. How could he know, when the person who bought the system didn't know these things existed, and therefore could not inform him?

Because turnkey procurement is not appropriate to advanced manufacturing technology project management does not mean that all process and equipment development must be done by the individual manufacturing company, in its own shops and laboratories. It does mean, however, that even where technology is procured from the outside, its selection and integration into the company's operations must be managed as a development project following the basic procedure shown in Figure 1.4.

INSIGHT, CONTINUITY, AND MOMENTUM

There has been recognition in the literature recently of the need for some one person to push a program relentlessly if it is to be a success.[2] It has been noted that if this so-called project champion is not present or is so low ranking as to be powerless, the chances of a program succeeding in a modern industrial organization are slight. Just having someone pushing a program, who meets some set of minimum criteria for the job, will not of itself assure success. Having a champion is no magic button that one presses to make all problems disappear.

There is never a magic button; there are just lots of problems, both technical and managerial, to be solved one at a time until none remain. The person who does this can be called a project champion, but he is really the principal investigator, or the project manager, and the things he is supposed to do are the things successful principal investigators and project managers have always done.

The first of these is acquiring technical insight. By living with the project daily, he develops a discernment about how the new manufacturing process works that is absolutely necessary to success. He also provides the other attributes necessary for success: continuity and momentum.

Every successful project has a logical path from original idea to the implemented production process. This path is never the route that seems obvious at the start, and it is usually not understood until the project is completed. It is unique, and finding it and getting from one end to the other as efficiently as possible is the second important function of the project manager. In order to do this, he must stay with the project, if possible, from beginning to end. No good project manager or principal investigator is ever willing to leave a worthwhile project unless it is completed. However, there is a tendency for some overly compartmentalized management structures to pass projects down through a bucket brigade of departments and project managers. Any technique which keeps at least one person from becoming deeply involved and thus developing profound understanding of the project must be avoided.

Momentum is the other attribute of successful projects which the project manager must maintain. Long pauses between phases while awaiting approvals and funding, or periods when no resources are available for other reasons, greatly reduce the chances of success. All programs, no matter how technically promising, have a funding and activity threshold below which no progress can be made. Any resources committed below this level are wasted. Many examples exist of these on-again, off-again jobs where someone who is free for the moment is put on the job and then pulled off. If the man-tech manager has neither the resources nor the interest to do the project, he must have the nerve to kill it and go on to something that will be successfully completed.

Maintaining project momentum and agility in ponderous organizations, which always seem about to lumber to a halt, is one of the most difficult tasks in man-tech development. One constantly hears the complaint that if any task—fixing a machine, buying a part, hiring a technician—is allowed to proceed through most corporate systems at the system's own pace, it will never get done. This is literally true, because a task which is done so slowly that when completed it is no longer relevant or useful was, from the project's perspective, never done.

The development and implementation of improved manufacturing processes and equipment is a race where the life of the company is at stake. Running the race in heats, with each heat being a separate project, does not make the results of eventual loss any less fatal. American industry, because of the massive lead we inherited from our fathers, became convinced that there was no competition and therefore no race. Anyone who still thinks there is no race is amazingly blind to the corpses of the losers all about him.

To eventually implement a new process and some day make a new product is not enough. Man-tech projects must be done with relentless urgency if they are to have an effect on the company's competitive position; otherwise they are irrelevant and useless.

Advanced manufacturing technology projects are sometimes conducted intermittently, because they are considered luxuries that can immediately be dispensed with if business slows. They are not luxuries; they are a legitimate cost of staying in business, the same as paying employees every

Friday or the electric bill every month. When in desperate condition, a company can postpone a payday or a utility bill, but eventually these must be paid or the company dies. Although the costs of developing and implementing the new technology which keeps a company competitive do not come due on a weekly or monthly basis, they too must be paid if the company is to survive.

If these legitimate costs are postponed too long and a company or industry must catch up quickly or die, it usually dies, because the technical and managerial skills needed to accomplish this difficult function no longer exist. The people with these skills will have left the company, and those who remain will not believe that management's new infatuation with high technology is anything more than a fad or a sham.

Desperate attempts to buy technology without the technical or management skills and insight needed to select, develop, and implement it do not solve a company's manufacturing technology problems. It is only a more expensive way to go out of business. The individual manufacturing technology project must be conducted with continuity and momentum. It is even more important, however, that the company have an overall commitment to improving its manufacturing technology that has these same attributes.

REFERENCES

1. Miller, J. A. From lab to shop floor—managing advanced manufacturing technology. *Mechanical Engineering,* **105**(11):62–68 (1983).
2. Peters, Thomas J., and Waterman, Robert H. *In Search of Excellence,* New York: Warner Books, 1982.

2
CORPORATE POLICIES AND STRUCTURE AND THE MANAGEMENT OF ADVANCED MANUFACTURING TECHNOLOGY

TECHNICAL SUCCESS AND MANAGEMENT FAILURE: THE TAIL ROTOR DRIVE SHAFT PROJECT

The first important man-tech project I was involved in failed. As a junior welding engineer, in the early 1960s, I worked in the materials and process laboratory of a firm that built helicopters. This department provided production support on existing processes and conducted, on a very limited basis, due to lack of resources, new process development.

Most of our time was spent supporting production, and the worst continuing production problem was the induction silver brazing of the tail rotor drive shaft which supplied power to the tail rotor from the helicopter's main gearbox. This assembly was comprised of a length of stainless steel tube and two end fittings which were joined to it (Figure 2.1). It was a nonredundant, rotating, safety-of-flight component. Its failure could cause the loss of the aircraft.

There had never been a service failure, but the brazing process was very sensitive to operator technique, and the brazed joints were very difficult to inspect with the methods then available.

To make the assembly, the operator first scrupulously cleaned the shaft, the end fitting, and a silver alloy filler ring. In brazing, only the filler material is melted, and this filler ring fitted into a groove cut into the end fitting. Both the tube and the fitting were next liberally coated with a paste flux whose function was to prevent surface oxidation during the brazing cycle. The melted silver flowed into the gap between the tube and fitting where it froze when the heat source was turned off, making the joint.

The heat source was an electrical induction coil into which the tube assembly was placed. A high-frequency electric current source was connected to the coil, and when it was turned on, it caused the shaft assembly to be heated very rapidly in the region in or directly adjacent to the coil.

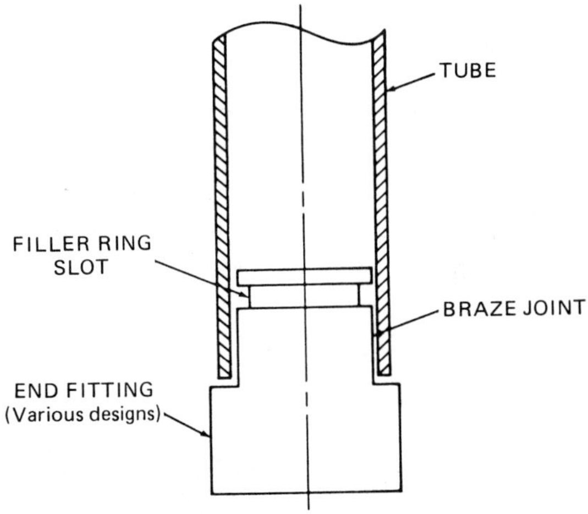

Figure 2.1. Brazes Shaft Assembly.

Once the brazed assembly had cooled, the flux was dissolved from the surfaces with hot water and the joint was X-rayed to assure that it was essentially filled with silver. In practice some small percentage of the total area was allowed to be unbrazed and these areas would show on the X-ray film as relatively darker areas compared to those where silver was present. If the joint was found acceptable, it was subjected to a static torque test. The part was then finished by applying a corrosion-protective coating to the end fittings.

If the brazed joint was rejected by X-ray, as often happened, the shaft was reheated in the induction coil to melt the silver and disassembled. The silver was then removed chemically from the details, and the joining process was repeated, often with no more success. The process was extremely sensitive to minor contamination of any of the surfaces or materials involved, to the gap between the tube and the end fitting, and to the heat-up rate and peak temperature.

The X-ray technique detected only the presence of silver in the joint. It provided no information about the quality of the metallurgical bond between the filler and the joint faces upon which the strength of the joint was dependent. The proof test was supposed to assure joint strength.

X-ray film interpretation was found to be operator dependent when the end fitting fell out of a shaft which had passed all inspections. This happened in a hot-water tank during the application of the corrosion coating. It was found that the brazing operator had neglected to put the silver ring in the groove before brazing the part. The flux had melted and then frozen in the joint forming a bond of sorts. The flux was of such low density as to be invisible on the X-ray film, but because the X-ray reader saw no voids, he inferred that the joint was 100-percent brazed when it was actually 100-percent void. The fused-salt "joint" was strong enough to pass the static proof test, which gives a good idea of the value of the test. When the salt was dissolved by the hot water, the assembly fell apart.

This was the existing technology which we were determined to replace. It was slow, expensive, and, because of the many inherent opportunities for error, very worrisome at best.

A program to develop a flash butt welded tail rotor drive shaft was conducted. I, as the junior engineer, did much of the work but was involved in the program management only as a more and more bewildered observer.

The company owned a large flash butt welding machine which was used only a few hours per week to make landing gear components. The service experience with these had been excellent. This completely automated process was so fast that in a small-batch type of manufacturing facility, such as an airframe business, it could never be used full time. The program, if successful, would provide a better, lighter shaft at much lower cost and eliminate the quality problems caused by the operator sensitivity and difficulty of inspection of silver brazing. It would involve no capital cost because the underused equipment was already in place, and there would be no production implementation problems for the same reason. After 25 years, it is still difficult to imagine a program with fewer obstacles to success.

In flash butt welding, two components with similar end configurations are butted lightly together in the automated machine which performs the welding operation. There is no filler material; the joint is formed from the base materials themselves. An electric current flows across the two detail parts, and because the resistance is highest at the interface, heat is generated there. As the faces become plastic with heating, the machine automatically increases the pressure between them, extruding the metal of the original faces and producing a strong metallurgical bond between clean subsurface material. Once the programmed heat and pressure cycle is completed, the machine holds the welded part through a controlled cool cycle and the process is complete. The process is very fast, taking only a few seconds, and because it is completely automated, it produces the same joint every time. After welding, the extruded materials are machined off, and the assembly is then completed by corrosion coating (Figure 2.2).

The process was controlled by welding three test specimens with joints identical to the part. One was welded at the beginning, one in the middle, and one at the end of each production run. These were then machined and pulled to destruction in a tensile machine. Failure had to occur in the tubes, away from the welds. If all three specimens for the part lot were

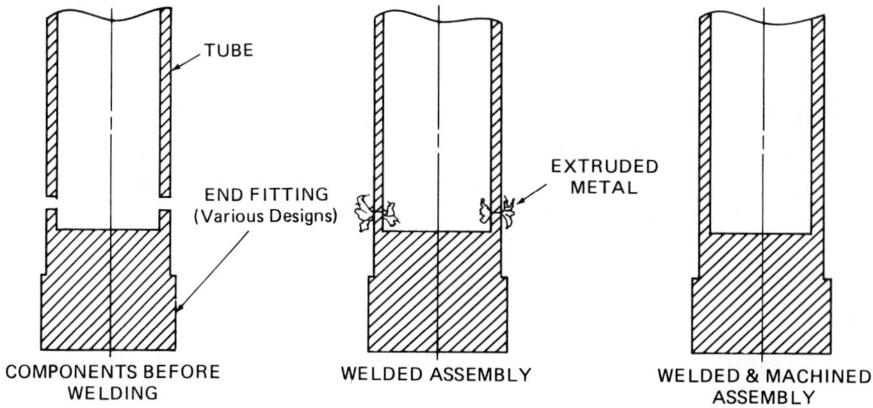

Figure 2.2. Welded Drive Shaft Assembly.

accepted, the lot was accepted. In practice, once the welding cycle was developed, there was never a rejection. The individual parts were inspected by a fluorescent penetrant technique for defects. Again, in practice the reject rate was zero.

First the welding technique was developed and shown to consistently produce welds meeting the inspection requirements. Because of the consequences of a service failure, the new shaft was very thoroughly tested. A complete metallurgical evaluation was done, then many welds were tested to destruction in a tensile test machine. All failed in the tube, away from the welds.

Next, several shafts were twisted to failure in a specially built rig. Again, all failed in the tube material away from the joint. While these tests were under way, tests were conducted to assure that a flash butt welded assembly was not more susceptible to corrosion than the brazed one. It wasn't.

The final hurdle was fatigue testing. Thirty shafts were made to be tested in a rotating beam machine. This subjected them to the worst type of vibratory loading seen in service. Twenty-seven shafts ran flawlessly, either not failing after millions of cycles, or, when run at higher loads, failing not at the welds, but in the tube material away from the weld. Because the tube material was the same for both the new and old assembly, this proved that the new was at least as good as the very best of shafts made to the old design.

Then came fatigue test number 28. The shaft was nearing the number of cycles at which it would have been retired without failure, when the testing machine broke. The flywheel shaft failed at full speed, throwing the test shaft, with a heavy flywheel still attached, against a support beam. The support beam happened to be directly in line with the weld, so that was where the shaft sheared.

Test 28 was discarded because the circumstances of the accident assured that the shaft would break at the location where it hit the support beam. No one involved with the program was concerned. We repaired the test machine and ran off the two remaining specimens. These, too, were successful.

The aircraft into which the new shaft was to go was used in both military and commercial service. The necessary reports were prepared for the Department of Defense and the Federal Aviation Agency. Fatigue test number 28 was mentioned, parenthetically, as accidentally destroyed in a test machine failure.

The reports were reviewed and approved by the government, and now the only step remaining was to change the engineering drawing and begin production. The process by which the drawing was changed was the same used in all aircraft companies; representatives of the various engineering functions sign off to show agreement. All of the groups—design, metallurgy, stress analysis, manufacturing engineering, functional test—signed, except one. The manager of an engineering project group disapproved of the welded shaft, and that is where it stopped. Attempts to overcome his objections went on for many weeks, slowly working up the hierarchy but never reaching anyone willing to reconcile the conflict. All of these meetings ended the same way; the objecting manager would point his finger at the project manager and say, "Admit it. A shaft broke in the weld. If it happened once, it could happen again."

Throughout this final stage of the dying project, the program manager kept his temper and patiently explained specimen number 28 and the broken flywheel shaft again and again. With each repetition, as he became more defensive, it sounded more like an excuse.

The program, of course, became controversial, and that was ultimately what killed it. In that organization, at that time, very few people wanted to be associated with a controversial change in a critical flight-safety component. The flash butt welded tail rotor drive shaft was never implemented.

The program's failure obviously had nothing to do with the technical merits of the innovation. The reasons for it were much more complex than just one person who did not know the difference between sound, conservative engineering judgment and what can only be called cowardice. He was, after all, the person the organization chose to put into the position and it was he who the organization allowed to make the final decision, if by default.

CORPORATE STRUCTURE VERSUS INNOVATION

The failure of the tail rotor drive shaft program was a failure of the company's management. Unfortunately, very few management failures are ever admitted, much less investigated. Technical failures are acceptable. No one can be blamed if he is thwarted by the laws of the universe. If he is thwarted by his boss's incompetence, it is better to blame mother nature. She does not fill out personnel appraisals.

Specimen number 28 was the basis for proving that the program was a technical failure, because it was the only thing in an otherwise perfect program upon which to base such a conclusion. To understand why intelligent people would accept such an unintelligent argument, we must understand why the project's failure was so necessary to the organization.

People need some structure upon which to hang their lives. In a manufacturing company a structure is necessary to assign specific tasks to specific groups who, it is hoped, will find the best ways to do them, and who are responsible for the performance or nonperformance of these tasks. But the structure soon becomes more than just a way to decide who buys material, who bends pipe, and who makes the soup for lunch. Business is, in spite of recent court cases, still dictatorial. People can be terminated without recourse because of lack of work and for other reasons. Termination is one of the most traumatic personal disruptions that can occur in this society. An important function of corporate structure is to provide some assurance that termination will not be whimsical; that there are rules which, if followed, protect one against being fired. Thus any corporate structure, and each person's place in it, soon becomes sacred even to those who spend most of their time complaining about the specific policies and assignments which comprise it.

Any change in the way tasks are done threatens to change the way they are assigned, and changes in the assignment of tasks threaten the structure upon which people have based their lives. That is a primary reason why the implementation of new manufacturing techniques is resisted, and why improvements which their developers think elegant and fascinating are frightening and even repulsive to many others in the company.

Obviously a structure is necessary. As the company becomes larger, the structure becomes more complex and more rigid, not so much because the company requires it, but because its employees need a more defined and complex structure as the company becomes more impersonal and more potentially whimsical in its treatment of them.

This structure is based horizontally on functional units which are assigned some specific group of tasks for which they are responsible. It is based vertically on the premise that those with more knowledge of the company's business are placed higher in its structure and are given more authority. Both of these concepts, unless carefully moderated, are very disruptive of innovation in advanced manufacturing technology.

Innovation and the Horizontal Structure

The horizontal division of tasks between departments can easily become a system of rules which permit only persons in certain departments to have ideas about certain subjects. The surest sign that someone has failed in his job is the necessity for someone else, from the outside, to be brought in to do his assigned task or correct errors. Unfortunately, this sorry state of affairs is confused with the totally different case of new ideas about the job. Often, because a person has been doing something diligently and competently, he is mentally blocked from seeing how to make large improvements in it. But an outsider coming up with a new idea is seen, not as a gain to the company, but as a failure of the individual most involved with the area of the idea. The new idea is thus transformed from a valuable aid to the person or department responsible for that area to a threat to their position and an insult to their competence. That is why, no matter how carefully suggested, new ideas are often met with the same emotional responses as threats and insults.

This was probably the reason the tail rotor drive shaft innovation was resisted by the only means possible to resist it: specimen number 28. The idea for the improved shaft came, not from the department responsible for the design and improvement of the power train, but from the welding laboratory whose duty was to find the best way to make the designs given to it. The department responsible for producing those designs was headed by the manager who fought and ultimately defeated the innovation. This was not a coincidence.

A person or group has an area of responsibility which they feel they have been discharging competently, with great dedication. When an idea about that area comes from outside the group, from someone who they feel has no competence or responsibility in this technical field, they (we) behave very badly. Any pretext is used to discredit the idea and those responsible for it. The violence of the reaction increases directly with the value of the idea and inversely with the status of the suggestor. People who are quick to recognize this "not-invented-here syndrome" in others, and despise it, will show the same symptoms when an outsider makes a suggestion about their assigned area of responsibility.

If this behavior is tolerated in a company, only ideas which fit neatly into one of the existing departments of the company, and originate within that department, have any chance of being implemented. Ideas that straddle functional units do not survive because they quickly become casualties of the border wars which flourish in such companies.

Yet, the dark and bloody ground at the borders of existing areas of responsibility is often the most fertile area for ideas. The ground within the existing fields of responsibility has usually been plowed over so many times that any ideas which are produced, although readily accepted and implemented, are trivial. The important innovations, like the welded drive shaft, straddle boundaries. The shaft not only solved manufacturing and quality control problems but, because it was lighter and inherently more dependable, offered design advantages as well.

The normal operations of companies and the way they select people for promotion aggravates this problem. People are promoted for their skills at accomplishing the tasks assigned to them while avoiding conflict with those assigned to other tasks. One does his job and does not interfere with those doing other jobs. This is an excellent way to operate an assembly line where rigidly defined and separate tasks must be done again and again. Those who succeed are those who never innovate and never interfere with others.

If these criteria are the only ones used for deciding the value of an employee to the company, and hence whether he should be promoted, a company's ability to accept and implement innovations may be destroyed. If someone whose duties involve evaluating, accepting, and developing ideas, no matter what the source, never makes a mistake, never crosses the boundaries which surround his department's tasks, and never becomes involved in a controversy or conflict, he is an anti-innovative drag on the company's progress. In order to defeat the drive shaft innovation in that company at that time, it was not necessary to rationally argue its technical merits. All that was necessary was to find some pretext to make it controversial. This was because so many of those involved had reached their positions in the company by never having an error traced to them and never becoming involved in a controversy. Skill at evading conflict was far more important than skill at reconciling it. This technique for success can be described as wearing a gray hat and riding a short horse in the middle of the posse.

Such managers will never reverse their defensive philosophy of corporate life because top management has endorsed it by promoting them, and they are usually pathologically sensitive to the wishes (real or imagined) of top management. It was not specimen number 28 that killed the new drive shaft, it was because specimen 28 provided a pretext for making it controversial, and the unwritten rules of the company said that anything that caused controversy must be buried immediately, supposedly for the good of the company.

For the true good of the company, management must not shrink from controversy or conflict, particularly if it appears to be the result of the not-invented-here syndrome or any other departmental border psychosis. When these occur, active steps to force the argument into the light, to find out exactly who the protagonists truly are, to define the issues in technical, not territorial, terms, and to reach the most technically sound solution must be taken. Extraneous arguments, such as questions of whose job such innovations are, who thought of the idea first, whose academic or professional background lends more credibility to his side of the argument, or whose name or what specific information should go into the patent application, must be ruthlessly excluded from such discussions. If they are not, no conclusion having any competitive value to the company will ever be reached, and lifelong hatreds between people who must cooperate for the good of the company may be created.

If upper management had done this in the case of the tail rotor drive shaft, either specimen 28 would have been discarded and the shaft implemented, or, if the protesting manager could produce any valid reasons for his often repeated fears, some additional tests could have been devised and conducted to see if those reasons were valid.

Perhaps, if such tests had been run and yielded evidence of an inherent defect in the innovation, it would have been possible to modify the design or the manufacturing process to solve the problem. But to spend the money and time to develop and thoroughly test a new product and then to allow it to be abandoned because it became controversial, without trying to reconcile the controversy, was wasteful of both the company's resources and a significant technical opportunity.

Dealing with controversy about innovations rather than running from them is important not only in terms of the particular innovation, but to evaluate the true contribution of the people involved to the long-term advancement of the company. Those who are the worst barriers to innovation are often those who are most glib with the latest buzz words about encouraging it. They may also be very adroit at disparaging ideas from other people, which, after a suitable aging period, they then produce as their own. Watching people evaluate innovations, particularly those that are controversial and come from others, is extremely important in selecting those who will contribute most to the long-term survival of the company.

Although these things are more commonly the responsibility of upper-level management, all levels must do everything they can to resist the not-invented-here syndrome and assure that they do nothing which encourages its growth. The most insidious is the temptation to imply to a subordinate who has been battling to keep an old process running that if he had been doing his job, he would have thought of something which someone else suggested, to solve his biggest problem. He may not have understood the true nature of this problem, because he was so occupied with saving the company from its short-term consequences. Rather, he should be complimented on his efforts in a way that implies that they are appreciated and that the new idea's implementation is going to need his help in order to succeed.

A Short, True Horror Story

A large shipping company received a contract to supply capacity on short notice to an important customer. In effect, they were given a retainer in exchange for their agreement to have vessels available where and when needed. The company president and the vice president of administration met and worked out a plan for providing the additional communications links made necessary by the contract.

Their solution required the procurement and installation of new equipment on every ship. The V.P. of administration made several phone calls from the president's office to suppliers and found that this would take at least six months. After it was decided that this was unfortunate but unavoidable, the manager of communications who reported to the V.P. of administration was called to the president's office to be briefed on the assignment, which would be his responsibility.

The assignment was explained, and then the manager, who was cursed with insight into the communications system and a compulsion to innovate, which he should have been smart enough to stifle, made a suggestion. He explained how by reprogramming the existing equipment he could free the necessary channels, making additional equipment unnecessary. He was sure he could have the present system doing exactly what was required in less than a week.

The president of the company was ecstatic. He was able to tell the customer that they would be able to start fulfilling the contract almost immediately. He complimented the manager. The V.P. of administration was silent.

Almost immediately the reason for his silence was revealed. He rationalized his envy of the manager for suggesting a better solution than his in the presence of the president as disloyalty. It was made very clear, very quickly, that the manager had no future with the company, now that his boss perceived his competence as a personal threat.

The manager has since left the company, but the V.P. survives. This is hardly surprising since personal survival was his only goal.

Innovation and the Vertical Structure

The horizontal structure of a company can define which departments people must work in if they are to have ideas about certain subjects. The vertical structure can become a set of rules about who, within each department, is permitted to have what innovative ideas. In its most perverse form, this can approach the situation where low-level employees are only allowed to have ideas of small value to the company, and ideas of increasing value must come from progressively higher levels of management. In this situation, any employee who conceives of an improvement more valuable than his station permits is treated as a renegade.

This is extremely unfortunate, because the factors which produce ideas for useful improvements are practically never found at higher corporate levels. These factors include profound insight into the product and the currently used processes for making it, and frequent and informal communication between those who have insight about various but overlapping aspects of the manufacturing process. Companies usually have all of the ideas needed to vastly improve their operations available to them at the lower levels of the organization. Many consultants have done extremely well when called in by a desperate upper management to solve seemingly impossible problems, by simply asking first-level supervision, engineering support personnel, and labor what should be done. Ideas which upper management would not dream of listening to from their own employees become gospel when reported through a thousand-dollar-a-day consultant.

Another result of this vertical structure problem is that those who have the authority to commit resources to their own ideas often have been away from the day-to-day operation of the manufacturing plant for so long that they have lost the necessary technical insight. This leads to selecting inappropriate technology or developing and implementing it in such a way that the integration with the existing shop and its operations fails. These vaguely wrong innovations force fed by upper management often progress far past the point where they should have been either killed or corrected. This happens because no one with an understanding of how to solve the problems has the authority to solve them or even to discuss them with those who do.

Quality circles, or work teams, if properly organized and led, can help to overcome both the horizontal and vertical structural barriers to successful innovation. They must be organized across departmental boundaries, and their communication with upper management must be direct, not through people who think their assignment is to shield upper management from anything they might not wish to hear. A quality circle, organized within one department, and led and structured so that it never enters any area of potential conflict between management factions, will quickly become a useless pipeline for buzz words and propaganda. Oddly, although the lower corporate levels will not be fooled for a second, top management, because of its isolation, may not catch on to what is happening for quite a while.

INNOVATION AND CORPORATE CONTROL SYSTEMS

Another organizational impediment to the progress of advanced manufacturing technology is the uncontrolled growth of complex approval and financial control systems. As an organization grows, it becomes necessary to institute and enforce certain controls. The basic theory behind these regulations is, in practice, not nearly so important to developing and implementing change as is their number and the complexity of the path they force an innovation to follow.

Each approval and each form that must be completed is always in itself justifiable and necessary. It is very easy to justify the installation of a traffic light at every location where a child might cross a busy street. But if the installation of each new traffic light does not take into account the number and location of existing lights, and if no attempt is made to synchronize them as a system, eventually all traffic will stop permanently. In many companies traffic would have stopped completely long ago, except that people do not obey the lights anymore. The system is constantly circumvented with the tacit approval of management. If the system can be circumvented by a high-minded innovator for the good of the company, then it can be circumvented by someone whose motives are less noble—a thief, for example—and the system itself is only a useless encumbrance. The purpose of approval systems is to prevent waste or theft of resources, but unless some insight about the request exists, approval or disapproval is a sham and the approval process itself is a waste of resources. If someone is asked to accept some responsibility by signing for something he does not, and usually cannot, understand, he has the choice of being a rubber stamp or an impediment. Unfortunately, there is much more gratification and much less personal risk in being an impediment.

Useless approval systems grow for a number of reasons. The first is failure to differentiate between those who must be informed that something is to happen and those whose approval is required before it can happen. For example, the accounting department must be informed that a program to develop a new manufacturing method will be conducted so they may set up the necessary bookkeeping accounts to provide the program manager with the financial information he needs to control the program. But because a signature is required to prove that the head of accounting was informed, this is misunderstood to mean that he must approve before the program can commence. This leads to a frustrated engineer, trying to communicate with, or intimidate, a perplexed and increasingly churlish accountant.

In a large and compartmentalized company, if no rigid differentiation is maintained between approval and information-only signatures, the system will invariably become unworkable. The nontechnical people will see themselves as assaulted by mad zealots trying to force them to sign documents which, with their limited understanding of technology, look suspiciously like their own corporate death warrants. The innovative technical people will see themselves as besieged by cowardly, ignorant bean counters whose only function is to stop progress. The confusion and frustration is intensified if top management has approved and is pushing the program, yet all of the necessary resources must be justified over and over to various departments, each in the proper order, for the most obscure of territorial reasons. If one has never had to work with such a system, it is almost impossible to comprehend the time that is wasted and the interpersonal and interdepartmental conflict and contempt it can cause.

These systems grow for all of the usual bureaucratic reasons—empire building in particular—but they also grow because, when looked at by each department manager, his part of the monster looks like a useful servant. He can completely justify to himself and to anyone else his part of the procedure based on the mission of his own department. Because he does not define the mission of his own department (upper management does) he is very unlikely to modify the system except to "improve" it. Improvement almost never means simplification, because to simplify is to do less of what he thinks top management is paying him to do.

So the system grows, making simple tasks difficult. The development of advanced manufacturing technology is not simple, so the system makes it nearly impossible. A new idea cannot be tried to see if it might work without being approved. Nothing can be approved without proof that it will work. Good ideas are stillborn because people no longer try to get them approved, and those who do try are considered overly aggressive troublemakers and treated as such. Eventually they leave the company or stop trying to make improvements. Things which might help to generate ideas by bringing the people with insight about the company's manufacturing problems in contact with potential solutions—trade shows and professional society meetings—are never approved. If enough people have to approve, one of them is sure to consider such things boondoggles from which it is his duty to protect the company.

Middle management, the heads of the individual departments involved in the approval system, cannot and will not simplify and rationalize it. Top management only can do it, and in many companies top management does not even comprehend the implications of the system because they never have to use it. Their requests for funding or services or maintenance personnel go on a simplified priority path. A divisional vice president who can have his office redecorated twice a year merely by telling his administrative assistant to get it done while he is out of town for a week will be unaware of what torment a laboratory or manufacturing supervisor must go through to obtain a $20 replacement part for a desperately needed machine.

Top management must understand and control the approval systems which in turn control the organization. There should be a periodic methodical review of such systems conducted by someone with the authority to modify and simplify them on the spot. He should look at them to assure that the number of approvals required is held to the absolute minimum. Anything requiring five signatures will be done very slowly, and anything requiring eight probably won't be done at all, at least not in the form intended. Those who do not or cannot understand what is to be done and its implications for the company should not be required or permitted to approve or disapprove. If they must be informed, a separate sign-off sheet should be used which states that they are acknowledging only being informed and that they are responsible only for performing their required functions, not for the success of the project.

These reviews should be done frequently, perhaps every three years. The first review will be a massive job. Do not allow each department head to review his part of the system and make recommendations; this will only make it worse.

Rationalizing and simplifying a company's approval and control system is actually the process of making these systems meet the same cost/benefit justification criteria that they force all of the other functions within the company to meet. A justification and approval document for petty expenditures which costs $300 to process and, at most, can save $100 should be eliminated (as should anyone who insists on keeping it). Even if a company never conducted a man-tech program, keeping its control system under control and reducing the resources spent on it would produce important productivity gains for a relatively small investment.

The system by which a company allocates resources to manufacturing technology development, both for the overhead budgets of the departments involved and for specific projects, is the most important single interface between the company and man-tech development. It is at this interface that the company has its greatest effect on the success or failure of these projects. Their purpose is to develop and implement the best manufacturing process possible and implement it as quickly and economically as possible. Conducting the cheapest project possible, regardless of results, is obviously not a sensible goal; yet, some corporate cost control systems permit only such projects.

A true development project attempts to do something which has never been done before. In later chapters, ways of controlling and reducing the costs and the inherent risk of such activities are explained, as are techniques for obtaining the hard information needed to produce accurate cost estimates and justifications. But estimates are estimates, and the exact cost of a given task or project will not be known exactly until it is completed; furthermore, its value will not be known until its cost can be evaluated against the benefits the company finally derives from those costs. Attempting to project the absolute minimum budget for success is like trying to find how close to the edge of a cliff one can run before falling off; a small error in one direction is less important than a small error in the other.

Unfortunately, the compartmentalized structure of many companies leads to the existence of positions and departments whose primary function appears to be arbitrary reduction of budgets with no consideration of the cost/benefit realities involved and no responsibility for the ultimate results of those cuts upon the project. This usually leads project managers to intentionally pad budget proposals to allow for arbitrary cuts. Once this process begins, the entire cost estimating and budgeting procedure becomes a sham.

Solid, written justification of proposed project budgets should be required. But similar justification should also be required for reducing them. Someone wishing to reduce a well prepared, explained, and justified proposal should not be permitted to do so simply by writing "too much, reduce by 30%" in the margin. Company policy must require him to explain why he thinks the estimate is too high and force him to take responsibility for the ultimate effect of the reduction on the outcome of the project. Under the one-way system of budget negotiation used by many companies, the same person can arbitrarily reduce a project's resources at the start and then criticize the project manager at the finish for being over budget or technically unsuccessful.

TRADE SHOWS AND PROFESSIONAL SOCIETY MEETINGS

The often contradictory demands of a company's baronial horizontal structure, its vertical caste system, its chanted slogans about encouraging innovation and technical excellence, and its cost control cliches can usually be kept fairly isolated, except for the one time when they all collide head on; when someone asks to go to a trade show or professional society meeting. When this happens, either the happily ignored contradictions between these four areas of corporate myth must be faced, or some policy which will allow them to continue to be ignored must be concocted.

The policies that result can be rather strange. For example, employees may be allowed to attend, but only if they give a paper and they may attend only for the day on which they present their paper. Or, the policy may require that all papers be given by a single designated person whose supposedly great professional eminence will thus enhance the company's and the paper's prestige. Or, society meetings and trade shows may be treated as awards for loyal service, and each year one person from a department, or perhaps from several departments on a rotating basis, is selected to attend. Such policies are irrelevant to why companies should support and endorse professional society participation and attendance at trade shows. They only assure that what little money is spent is wasted and brings no advantage to either the company or the employee.

People should be sent to trade shows and professional society meetings first, to acquire technical knowledge which will enhance the company's competitive position, and second, to enhance their own and the company's professional reputations. There are no other reasons.

If the company views these essential activities as perquisites or boondoggles, the people to whom they are awarded will view them the same way. In the worst case, a senior executive attends with his spouse (or significant other) on an expense-paid vacation. This is clearly a misappropriation of company funds which good management will not permit. Those who attend should know that they are there to work, that they are expected to get the most out of the experience, and that upon their return they must report it in detail. Anyone who has ever attended such a meeting has seen the crowds of Japanese photographing, questioning, and aggressively working for their companies. This is a sad contrast to some of the vacationing Americans one sees only at the cocktail and dinner parties given by suppliers. The attitudes of both groups accurately reflect their companies' attitudes.

The purpose of trade show attendance is the generation of ideas. For an idea to happen, insight about the company's manufacturing operations and information about new technology must come together in one mind. Therefore, only those with the necessary insight should attend trade shows and meetings. This means that sending most middle- and upper-level managers is not nearly as productive as sending people from lower levels who have up-to-date, detailed knowledge of the company's operations and problems. If the company wishes to pass out managerial perks, send them to golf tournaments, but send the employees who will produce ideas to meetings and shows, and send them there to work.

If several people within the same department, or from different departments, who are involved with the same technology, can all make a sound case for attending the same meeting, send them all. If the cost controllers think this is extravagance, explain what the essential costs of remaining in business over the long term are, and then send them back to cut the bureaucratic fat out of the administrative departments.

A company should encourage its professional employees to present papers and write articles about their work, not for any altruistic reason, but quite simply because this attitude permits the company to attract and keep the best people. Some managers try to discourage such communication, justifying their position with trade secrecy even in instances where it is clearly not the case. These managers usually feel that any professional eminence their employees gain threatens their own position or their control over the employees. They are right, it does, and it is an excellent thing for the company if the position and control of such managers is threatened.

All technical papers should be written by those who actually did the work, and the person who writes the paper should always present it. If there are several authors, probably the most junior should give it. For a boss to present or publish a paper written by a subordinate about work the subordinate did is contemptible. It destroys the atmosphere of intellectual honesty and equality that is essential for the conduct of successful innovative work.

CORPORATE STRUCTURE AND THE MANUFACTURING TECHNOLOGY DEVELOPMENT DEPARTMENT

Where the department or subdepartment responsible for the development and implementation of advanced manufacturing technology is located in the corporate structure, and what additional duties it performs, will have great influence on its effectiveness. Dividing the development and implementation between two departments, with the transfer taking place at the door to the production shop, is generally a mistake. The two times when a project is most likely to fail are at its very inception and when entering production. Then, when the most resistance is met, is the very worst time to transfer responsibility for the project. This division is superficially attractive, because it divides program responsibility along the same lines as other functions are divided between departments. If such a structure is used, no one has responsibility for the project, because each group can blame the other for its failure.

If it is accepted that one group or department should have responsibility for the development and implementation of a given project, and that their responsibility should end only when the manufacturing process and its equipment are implemented in production, the next questions are, for what major department should they work, and what other functions, if any, should they perform?

The group which develops and implements advanced manufacturing technology must have three areas of insight in order to succeed. They must learn about the new technology, but it is equally important that they understand the company's products and the current processes for making them. They must also understand the manufacturing shop in which the innovation must eventually operate. This group's position in the company's structure must assure that they are constantly increasing their knowledge in all these areas.

The obvious conclusion is that the group should be a part of manufacturing and be involved in day-to-day production operations. This situation has the important drawback that if these people are knowledgeable, competent professionals, they will be so valuable as a fire brigade that they will never start, let alone finish, an advanced technology project. Manufacturing operations are always very short-term-oriented. They must be for the company to survive in the near term. A competent group, attached to manufacturing, will never be permitted to deal with the long-term solution to a problem, because these are the people best able to find a quick fix, somehow, today.

In the late 1960s, a turbine engine manufacturer had a major problem with the assembly of the power shaft and turbine disk in one of its engine models. Engine failures were found to be caused by metal fatigue of the disk from fretting inside the bolt holes where it was attached to the shaft. The approach of a short-term-oriented group would be to find a better way to drill more precise bolt holes and to make better bolts. Neither of these was the answer; making an all-welded shaft, and eliminating the bolted attachment completely, was. But it required developing new welding and heat-treating techniques and equipment in addition to completely redesigning the shaft/disk assembly (see Chapter 5). No manufacturing-based group would ever be allowed the time to do this. They would probably be so involved in bolt and bolt hole technology that they would never think of it.

If a company has a manufacturing technology group attached to manufacturing operations and it is not constantly called upon to help with problems, the staffing of the department and its relations with manufacturing supervision should be investigated. There is a real problem.

The man-tech group must have autonomy from operations management so they will not spend all of their time troubleshooting; yet, they must have the insight that can only be gained by troubleshooting the existing processes and making constant incremental improvements in them. They must work on the production floor, not just observe, in order to gain the respect of labor and supervision and be seen as a trusted part of that team. This close relationship is essential when a new manufacturing process is being implemented. A way to reconcile this contradiction is for the group which develops and implements new manufacturing technology to report to an engineering function, but also to provide technical support to the company's manufacturing and maintenance departments and subcontractors. But, however they are fitted into the structure, they must be able to develop the insight needed without losing the independence necessary to conduct development projects.

Another approach, which has been used with particular success in the computer industry, is setting up competing groups with overlapping or even duplicate functions.[1] This approach provides a competitive atmosphere which is said to improve innovative productivity. If the competitors are evenly matched and evaluated only on results by people who understand those results and cannot be bamboozled, this method can be very valuable. But it requires a certain corporate culture, and it is not a magic cure for a company with staffing and approval system problems. If misapplied to a weakly led, poorly organized company, it will probably make things worse by aggravating the cutthroat politics and rampant distrust that already flourish in such organizations.

The competition must be fair, or the result can be less innovative productivity than would result if only a single group existed. The most obvious form of unfairness is if one group, by its nonoverlapping functions, can impede its competitors' programs. For example, if there are two groups, but one of them processes all equipment procurement requests, the difficulties which their competitors might have in obtaining facilities are obvious.

The competition must also be fair in that both groups must be staffed with people of equal competence and drive. Once one group has a significant advantage in this area, the imbalance tends to grow rapidly, for the better people leave the weaker group, and good people join the more successful one. When this happens, management's best move is probably to declare the game over, disband the weaker group, and reinforce success.

There are a number of other ways of fitting a manufacturing technology department into a corporate structure. The important conditions are that placement of the department should fit the particular company and that the department should deal rationally with the conflict caused by needing to work closely with production without becoming enslaved to finding instant solutions to short-term problems.

Making the technical support of manufacturing operations independent of operations has advantages not only in allowing long-term solutions to be sought but in solving short-term problems as well. Day-to-day problems on existing, proven manufacturing processes should, theoretically, never happen. So when something does go wrong, it means that someone did not do his job perfectly. A machine was not maintained, incoming parts were not correct and not found by inspection, or some shortcut was tried and failed. It is usually much easier for someone with independence of the organization responsible to dispassionately decide the best solution and see to its implementation, because he can resist the pressure to hide or distort the problem for political reasons.

Another way to incorporate an advanced manufacturing technology department into a company's structure is to make it a separate staff function with no responsibility to manufacturing or to the departments that support manufacturing operations. In this way it is free to explore and develop new ways of doing things, without any preconceptions based on the old way of doing things. This sounds much better in theory than it works in practice.

A department operating under such a charter is not only without preconceptions, it is without the basic knowledge of the company's business it needs to do its job. Because it has no production support responsibility, it has no responsibility for long-term production implementation of its projects.

This leads to a department with no insight into the company's business, spending large amounts of money on projects for whose successful implementation it is not responsible. The usual result is turnkey programs in which inappropriate equipment is dumped on operations and production support people who had no say in its procurement.

For the purpose of this book, it will be assumed that the man-tech department is part of some engineering function, separate from manufacturing operations but providing substantial production support on a daily basis.

STRUCTURE OF A MANUFACTURING TECHNOLOGY DEPARTMENT

In order for the manufacturing technology department to perform its functions well, it must be able to seek large improvements in manufacturing processes, yet it must be involved in the day-to-day support of the existing ones in order to gain the insight needed to find and implement the big jump. It must operate under control with its objectives and plans defined and understood, yet it must be flexible and have an atmosphere of almost anarchic intellectual freedom.

How does one reconcile these contradictions? First, by admitting them and understanding that the structure of such a department must be different in concept—not just in the way the boxes are drawn on the chart—from most other departments in the company. Most departments have fairly standard tasks which are performed over and over again, but this department is a machine for generating change; it may never do the same thing twice in precisely the same way.

It is, of course, impossible to design a manufacturing technology department which will fit exactly into every company in every industry. The department I will describe is obviously one primarily suitable for a fairly large aerospace company, because that is where I have gained much of my experience. Several other companies in other industries have evolved almost identical structures. The basic problems and their solutions are common, however, to many if not most companies, even if the actual skills needed and specific products produced vary. Tradition and corporate history have great impact upon the techniques various companies have developed for solving problems. It is only recently that discussions of varying corporate cultures have become fashionable. These cultures have always had great impact on what methods, organizational structures, and management styles were socially acceptable and would work in a given company. The department described here is not intended to become the next management fad and to be implanted, intact, no matter how inappropriate, into every company. It is intended to demonstrate the way things are intended to happen, are encouraged to happen, and sometimes actually do happen in a successful manufacturing technology department.

The department will have a number of disciplines for which it is responsible: for example, machining, welding, thermal treatments, chemical treatments, and coatings. It will also have various functional responsibilities: new process development, production support, customer service, and subcontractor approval and support. In addition, there will be development programs in various stages, ranging from ideas that are just hatching to completed implementations that are being improved upon or for which additional facilities are being bought, built, and installed.

These things must somehow be divided between subdepartments or individuals, but this division cannot be done within the rigid borders used in defining assembly line tasks. There are very few places except on assembly lines where that method of task assignment works, because so many things fall between tasks. Yet, if responsibility is not defined, things also fall between assignments.

The first step in dealing with all these contradictions is to realize that it is impossible to set up an organization which will run like an assembly line. There are so many overlaps and gaps in any planned structure for such a department that, at best, it is going to require a lot of communication between individuals and a lot of work by supervision and management to assure that the required communication takes place.

People have to talk to each other in this business, because it is not a production line were everyone knows where his job ends and the next begins. It is a joint creative effort where things are constantly changing. They must, because change is the primary product of such a department. An automatic machine which will run in icy silence to manufacture change cannot be built. The best one can get is a roughly assembled jury rig that is perpetually changing itself and is kept operating by each of its parts constantly exchanging information with all of its other parts.

Figure 2.3 shows the functions which such a department should perform. Exactly what the marketing and customer support functions will be will depend on the particular industry and who the customer is. But these are the essentially interrelated functions which develop process and product insight at the same time that they use it to innovate. The next step in defining an organization is to draw boxes around these functions and then place people in the boxes to perform the functions. This process of putting functions and people into boxes has become so automatic that we do it without realizing its negative implications, or that it is not necessary or

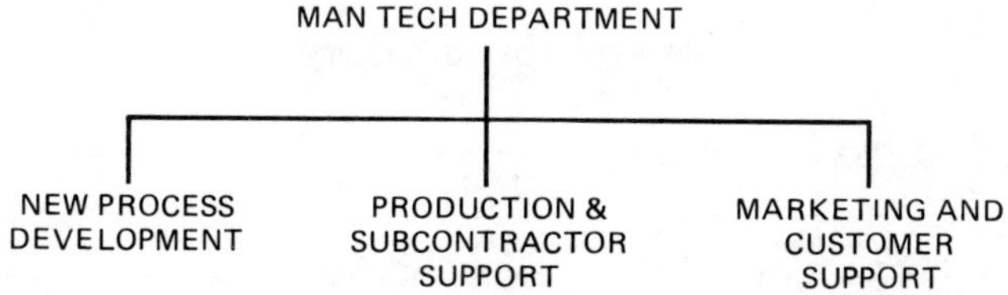

Figure 2.3. Organization of Manufacturing Technology Department by Function.

even very smart in many management situations. A manufacturing technology department is definitely such a case.

Another way of dividing the tasks would be by manufacturing operations. For example, all functions involving machining, whether new process, development, production support, or subcontractor support, are done by one person or group. All chemical processing functions are done by another, and so on (Figure 2.4). This leads to one component or product having a manufacturing process for which no one has overall responsibility because different people are responsible for each operation.

A method that supposedly deals with the confusion which results from trying to have both organization by function and organization by discipline simultaneously, while compulsively assigning people, assembly line style, to boxes, is matrix management. This, at its worst, tells a project manager that he has only the responsibility for progress; all of the resources he needs are locked up in little boxes which someone else controls. It is up to him to cajole, charm, intimidate, scrounge, or do whatever else is necessary to get the job done.

If practiced in its primitive, brutal form, matrix management generates emotional conflict and great frustration. Of all the things which will assure that a person does not reason logically or produce useful ideas, and will in the end spend his time hating his work rather than doing it well, emotional conflict and frustration are the surest. If a person is angry, or feels trapped or belittled, or if management expects extraordinary performance from him while avoiding their own responsibilities to make his job at least possible, he is not going to invent or implement anything. Perhaps a person will lay bricks faster if he is angry and frustrated, but he certainly will not do work requiring profound thought. He may look like it; he may be at his desk scratching a piece of paper or out in the shop wandering around, but no one who is angry can really think of anything else. Whatever department structure is adopted, it at least should not actively prevent people from doing their jobs, even if it does not help much.

Matrix management in a form so benign as to be almost unrecognizable is probably the best way to organize a man-tech department. The major

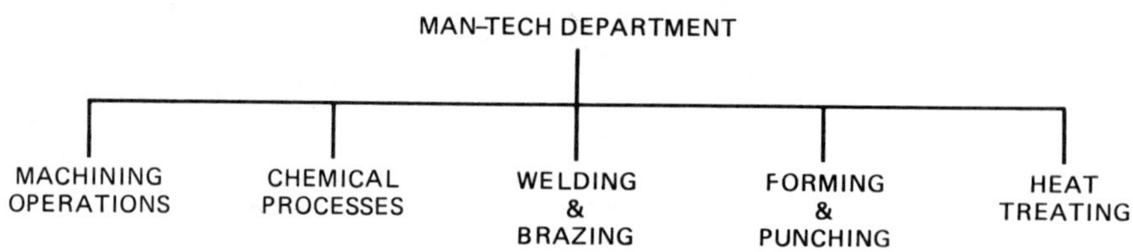

Figure 2.4. Organization of Manufacturing Technology Department by Manufacturing Process.

functions of the department (Figure 2.4) are each the responsibility of one individual: a supervisor of process development, production support, or whatever. He has overall responsibility for the administrative functions required, such as project plans, task assignments, and reports. He concentrates most of his time working within the area he is supposed to be supervising; but supervising is not actually his function. He is the leader of those involved in the function when they are actually working in it. He should be selected because if no one were formally assigned to lead, he would do it anyway by mutual consent. The "supervisory" part of the job, the paper shuffling and pencil sharpening, is not why he is selected; leadership based on mutual respect is.

These supervisors are the only people assigned to any specific functional blocks on the organizational chart for other than administrative purposes. These administrative assignments are based on where it is expected a particular person may spend most of his time and on how many desks there are in each office. Any attempt to set up more definite boundaries between functions is bound to be a failure in a successful department. It will, however, succeed in a failing department. Even the actual work done by the supervisors will cross the functional boundaries, and the supervisor of development will spend time working in production support, because he happens to have technical skills needed on a particular problem, and the supervisor of production support will undoubtedly have to help on development projects where his insight is needed.

For example, the supervisor of development may be the department's computer wiz, and if a tough programming problem comes up in production or anywhere else, he will be the one who has to solve it. Too often, the really skilled people are not the ones who become the group supervisors because of the mistaken idea that management should only manage. It is not uncommon to hear of someone turned down for promotion because he was too valuable in his technical functions, and those talents would somehow keep him from becoming an effective manager (whatever that is).

This unfortunate attitude results from not knowing the difference between management, which is essentially leadership by example, and administration, which is essentially clerical. This brings us to another key position in this department: the Secretary. The group may be large enough that it may have more than one clerical person, but there is only one Secretary with a capital S.* She is the absolute boss in administrative matters, and to the greatest extent possible, she frees up the technical

*I refer to the secretary as female because I am 50 years old and a child of my times. She may, in this context, be either male or female, as is the pronoun *he* throughout this book.

staff to deal with the technology. Get the best person that can be found and pay her more than she, or the personnel department, think she should make. She will be the equivalent of hiring a top engineer or scientist, because she will keep at least four top scientists or engineers from spending 25 percent of their time chasing travel vouchers and looking for lost requisitions.

Do not hire an "administrator," particularly one with a business administration degree. Such people only create more paperwork than they process, because they are indoctrinated with the truly weird idea that one can manage something without understanding it if only those who do understand it fill out enough forms and write enough reports. Hire an intelligent, well trained secretary, or better yet, promote her, get her the best office automation equipment money can buy, and let her make the department free of the biggest single barrier that innovation faces: administrative red tape.

The structure of a department of manufacturing technology is thus one where a department manager has reporting to him supervisors of the various functions required and a secretary who frees everyone from administrative paper so they can play with hard, shiny things. The effort of the individual technical people in the department, while focused toward one or another of the functions, is not exclusive to that function. Each will have projects or problems for which, as a principal investigator, he is primarily responsible. As technical specialist in a particular discipline, everyone is involved in any program or problem where his skills are needed. By working closely together, they have gained enough insight in each other's specialties and enough team spirit, so that when one of them is overloaded, another naturally gives him a willing hand.

This mutual assistance pact is independent of rank. The manager and supervisors will work with, and take direction from, the principal investigator, not for any equalitarian delusions, but because he best knows what has to be done. He must; it is his project.

The specific technical disciplines that the staff of such a department will be trained in is, of course, dependent on the business in which the company is engaged. However, except in the most specialized companies, a mix of skills and backgrounds is essential. Too often a department manager will think that his education is the ideal and will try to staff with only those of similar backgrounds. Most manufacturing processes today involve so many diverse disciplines that no college major trains one for more than a small part of the specifics involved. A department staffed with all manufacturing engineers, or all mechanical engineers, or all metallurgists, will be severely handicapped compared to a department with one or two of each, in addition to a chemist, a computer scientist, and someone with an electronics background. Diversity within individuals is also valuable:

a mechanical engineer with graduate work in metallurgy, a chemist with graduate work in computer science.

This is not to imply that these people will work within their educational specialty all or even most of the time. The main thing they will have gained from their education is the basic mathematics, science, and reasoning skills needed to attack a problem. They should have minds that were broadened, not narrowed, by education. Their education will provide the basis upon which the essential insight will be built, but it will, in most cases, provide none of the specific knowledge needed. This will come only from day-to-day involvement in the design, manufacture, and service of the company's products, and from their own flexibility and curiosity.

THE M-FILL™ PROCESS: AN EASY SUCCESS BASED ON MANY FAILURES[2,3]

The repair of service damage to gas turbine engine high-temperature components is very difficult because of the materials from which they are made. These are usually nickel-based superalloys in which crack-free welds are difficult to make (Figure 2.5). The problem is complicated by the

Figure 2.5. Gas Turbine Engine Nozzle Assembly.

welding filler wire materials; those that give the most defect-free joints do not have very good high-temperature properties themselves and thus do not last long in service.

There are two types of defects requiring repair: cracks and holes. These defects result from the service conditions these components experience in the hot gas path of the engine.

Brazing appears to be a more promising way of repairing these defects because the base metal is not melted in brazing and therefore the cracking which results from melting the base metal in welding is avoided. Brazing of these superalloys is complicated because they form stable surface oxides which prevent wetting of the surface and good bonding by the brazing filler materials to the base metals. They can be brazed with special surface preparation, though. This preparation usually involves immersion chemical cleaning and a thin coating of nickel plate to keep the oxides from reforming during the brazing cycle. These techniques work well for joining separate detail parts which can be cleaned and plated prior to assembly. They do not work well for cracks or small holes.

A crack has an immense surface area compared to its volume, so even if chemical cleaning solutions can be made to flow into it, the amount is so small that the solution is depleted immediately and the surfaces are not cleaned. Small holes or ground-out cracks, on the other hand, while easy to clean and plate, were impossible to braze with then-existing filler materials because they did not bridge well and tended to flow into the interior of the part, when molten, rather than filling the holes.

In the summer of 1980, a customer returned two turbine nozzles from engines used in a marine application. The engines had been run on diesel fuel rather than jet fuel. This had resulted in the generation of carbon particles which had eroded holes in every one of the nozzle vanes.

The replacement of the vanes in these nozzles is very complicated and costly. The replacement of the vanes in these particular nozzles was not feasible because every vane had a hole. They were therefore considered scrap.

Several expensive parts being scrapped at the same time because no method for repairing them existed provided the impetus for trying, once again, to develop a repair technique. A brainstorming session was held at which all of the previous, unsuccessful approaches to the problem were reevaluated. There had been many attempts, both with welding and brazing. The conclusion was reached that the best way to make repairs would be with a brazing alloy with good high-temperature properties which would wet well at brazing temperatures compatible with the base metals, but would not flow and run at those temperatures. No such alloy then existed. Most brazing alloys are formulated to join two pieces of material together—to act like glue. We needed to develop an alloy for filling holes: a high-temperature putty. Being able to define the problem did not result from any stroke of genius; it came from having tried so many other approaches that had failed.

The problem of surface preparation was solved by adopting a commercial system for cleaning and nickel-plating local areas using a wet probe rather than by immersing the parts in tanks. This eliminated the complicated masking operations.

All that was needed was a metallic powder which could be mixed with an organic binder and applied to the holes. When the part was heated in a vacuum to about 1900°F, this material would melt, wet, and bond to the surfaces while bridging the holes and not flowing out onto the part surface or into the interior passages. When it had been applied and thermal cycled, it had to produce repaired surfaces which, with proper coatings, would survive in the turbine engine gas path.

The central technical questions to be answered were therefore, what was the formulation of a filler with the desired properties, and what thermal treatment would produce those properties? The details of surface preparation and protective coatings could be worked out based on existing techniques once these central technical questions were answered. They were attacked first.

Based on insight into the high-temperature brazing of nickel-based alloys, a basic formulation for the filler was suggested and several variations were agreed upon. These bracketed the possible combinations of chemistry. It was agreed to mix small samples of each and run them through a series of thermal cycles to try to find a combination meeting our requirements.

All of this took place in a single two-hour meeting. No approvals were necessary at this point in the program, because the manager of the department had a small account budgeted just for this sort of quick preliminary investigation.

One of the first formulations worked, and the thermal cycle was optimized almost as quickly. Within a week a very promising process had been developed. Within a month, based on successful test results, the new process had been used to salvage the two nozzles from the marine application.

The process has since been subjected to extensive testing, both with laboratory specimens and in engines. As more understanding and faith in the material was gained, it was used for more and more demanding service conditions and components. During the period of development, the biggest problem was with people who had heard about the material and wanted to use it on components for which it was not suitable. This attitude was in marked contrast to the implementation of the flash butt welded drive shaft, in a different company at a different time.

Today, the M-FILL process is protected by a number of patents and is licensed to companies specializing in turbine engine repair both in the United States and abroad. Millions of dollars worth of hardware have been repaired with it.

INNOVATION IN A MANUFACTURING TECHNOLOGY DEPARTMENT

The M-FILL process development, as described, appears deceptively easy. The program itself looked almost too easy to some observers at the time it was conducted. This is because I have not described, and they were not aware of, the many other techniques that were previously tried and found not to work. Those who sat down and mapped out the answer in that two-hour meeting had gained the insight needed for this success by their past technical failures. They were involved in production, subcontractor, and customer support. These activities, as well as professional society contacts, were all necessary prerequisites to knowing how, finally, to solve the problem.

The key to managing the frequent need for failures as a prerequisite to success is with the very simple funding approval system which permitted this program to be conducted. At the beginning of the fiscal year, a modest amount of money had been placed in an almost undefined project called "Development of Improved Manufacturing Techniques." Only the approval of the manager of manufacturing development was required to spend this money. It was intended for the quick exploitation of ideas that might come along during the year. Since he was one of the M-fill conspirators, the program was begun at once without having to be justified to anyone who was not involved. This was an absolute requirement for success, because with the long history of failures to find an answer to this problem, it is doubtful if the project would have even begun in a company with a formal and complex financial approval system. Somewhere in such systems there is always someone who must be absolutely guaranteed success before he will approve a project, and who will never sign up for a project with a history of failure.

To many people, particularly those with legal and financial backgrounds, setting up a fund controlled by one person, no matter how small and for what purpose, is very poor management practice. They are so convinced that every penny and every second of labor must be protected by adversary checks and balances, that just the existence of such a fund assures its misappropriation. Edmond Burke's statement, "Great empires and small minds go ill together,"[4] is probably the best answer to this objection. The idea that the company's money is better allocated by a committee of those without insight into the problem than by the one person who understands it best is in itself impossible to justify. This fund was not intended to conduct complete development programs, but to permit preliminary investigations of ideas to see if further development was justified. It would better have been called an "Incubator Fund." The development of the M-FILL process went so well and cost so little that it was entirely financed from this fund. Most of the testing involved engine running of repaired components, and these were piggybacked onto engine tests conducted by other programs.

The establishment of such a fund is an example of the intelligent management of opposing risks. The risk of misallocation of a few thousand dollars is far outweighed by the risks inherent in a system that prevents the prompt investigation of new ideas in a company in a technically competitive society. If management feels that the risk of misallocation or waste is so great as to be unacceptable, then the person designated to control the funds should be replaced with someone who is not such a risk.

The previous failures to find a solution to the problem of high-temperature component repair were a substantial help in gaining acceptance of the M-FILL process, because they had always been admitted. These earlier attempts had developed various techniques which were unsatisfactory because they did not consistently produce crack-free welds, or because the properties of the filler materials used gave very poor service life. There is always the temptation to proclaim at least a partial success in any project and put the process into use with various limitations and disclaimers. One of the most important criteria for an acceptable manufacturing process is that it must be controllable in a production environment, and none of the previous attempts had produced a controllable process even if, under very close supervision, they had produced an acceptable repair on a part meeting very limited, specific criteria. Maintaining those controls and assuring that only parts meeting those specific criteria were repaired was really not possible in an engine component overhaul environment. Therefore these processes were not implemented, even when there was considerable pressure to produce a success. This meant that when a dependable process was developed, the department had the credibility to have it readily accepted. The manufacturing technology department must be its own toughest critic; if it does not honestly evaluate and report its own project results, it will have great difficulty getting them accepted. The reputation that should be cultivated is one of never beginning implementation unless the process is truly production suitable, and then always succeeding.

ADVANCED MANUFACTURING TECHNOLOGY DEVELOPMENT LABORATORIES

There are basically two types of laboratory facilities necessary for the development of new applications of manufacturing technology. These are, first, the facilities needed to investigate and evaluate new processes and products, such as measuring equipment, metallographic equipment, facilities for chemical analysis, and facilities for conducting tests appropriate to the particular product. The second type of facilities are the machines and other devices used in the process being developed for the particular application. All of these do not have to be owned by the company and set up on its premises for successful development programs to be conducted.

The first type of facilities are standard equipment which a reasonably well equipped company will own even if it never conducts a man-tech program. How extensive these facilities are is dependent on the size of the company and the industry in which it is involved. Those specialized facilities which the company does not have will usually be available as a contracted service from commercial testing laboratories. Some minimum facilities must be available without having to go off site if this work is to be done easily and efficiently. Whether the services required—chemical analysis, for example—should be part of the man-tech department or be supplied by a separate chemistry lab, in a matrix project structure, or be a purchased service, will depend on how frequently such analysis is required and on how specialized and extensive it is. These, in turn, will depend on the nature of the process being investigated: for instance, an occasional check of incoming materials in a metal working project as opposed to the continual chemical analysis required in developing a petroleum reduction technique. Obviously, such laboratory facilities, organized in a manner suitable for the business and product and readily available to the man-tech department, are necessary to the conduct of development programs.

The second type of facilities, the machines and devices necessary to the improved manufacturing technique being developed, are normally considered the facilities of a man-tech development laboratory. Owning such facilities can help development programs by making the quick trial of ideas possible. For example, if a department has its own laser in its applications lab, many more ideas may be generated and tried than if each idea must be the subject of a purchase order to buy time on a machine for its trial. Doing this work in one's own laboratory also protects proprietary information far better than any nondisclosure agreement.

Just the presence and the day-to-day operation of such a machine will develop the process insight needed to find additional applications for it. This was the way I became involved with electron beam welding in the 1960s. The technology was just coming onto the market, and a machine was selected, procured, and set up in the applications lab. An independent research and development program was funded to investigate the welding of a number of common aerospace materials, and work was begun.

Actually, it began too early. The technology was not nearly as developed and the equipment not nearly so dependable as advertised. In the long run this proved to be an advantage. One can learn much more about a process from a machine that does not work too well than from a perfect one.

Three years after the first laboratory machine arrived, a production machine was bought and installed, and the first of many applications was successfully implemented. The original machine cost about $160,000 at a time when the prime lending rate was about 3 percent. It took almost a year to deliver, and it was another two years before it was debugged and the procedures for manufacturing the first two applications were developed and implemented.

If the same machine were built today for the first time, it would cost at least $800,000 with the cost of money perhaps 10 percent. The development and implementation would certainly be no faster, but their costs would be far higher. This slowness and high cost have made the normal applications laboratory approach to conducting manufacturing development obsolete in many cases.

Figure 2.6 illustrates the basic problem with this approach to program management. It can be assumed that, as a program proceeds, insight into

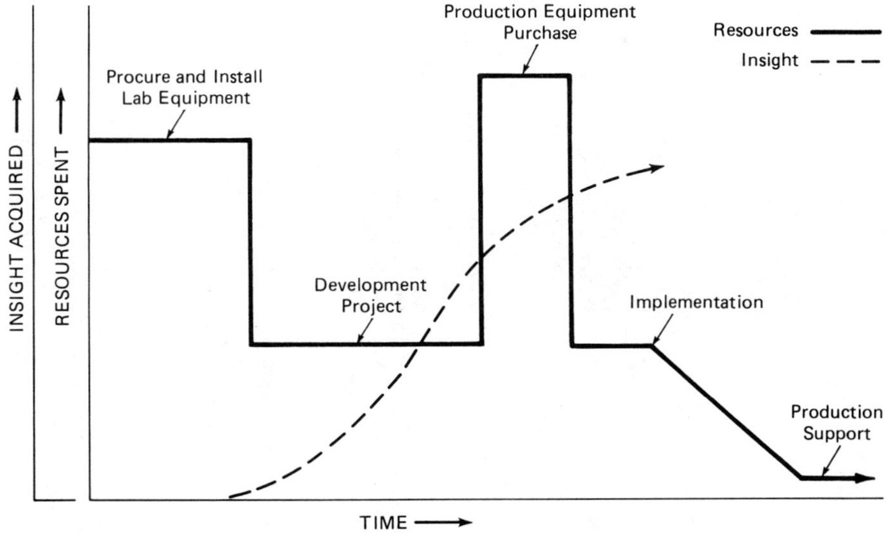

Figure 2.6. Resources and Insight vs. Time — Company Laboratory Approach.

the process will be acquired. This is, after all, the reason for conducting programs. In the applications lab approach, most of the money invested in a program comes at its very beginning, when the process is least understood. For this reason, pressures to begin programs with large capital equipment purchases should in most cases be resisted. One of the most important principles in the rational management of advanced manufacturing technology is to commit resources in proportion to how much is known about the process under development (Figure 2.7). Rushing to buy a machine for the lab is the most common violation of this principle.

This approach to program management increases the inherent risks by spending large sums on equipment early in the program, before the requirements the equipment must meet are understood. If one tours applications laboratories, one will often see, standing in corners, unused equipment, which shows that these uninformed guesses are often wrong. There are important advantages to be gained from failures which may eventually contribute to successes. But this does not mean that one should actively encourage failure or attempt to increase its costs.

Early purchase of laboratory facilities is required in cases where the

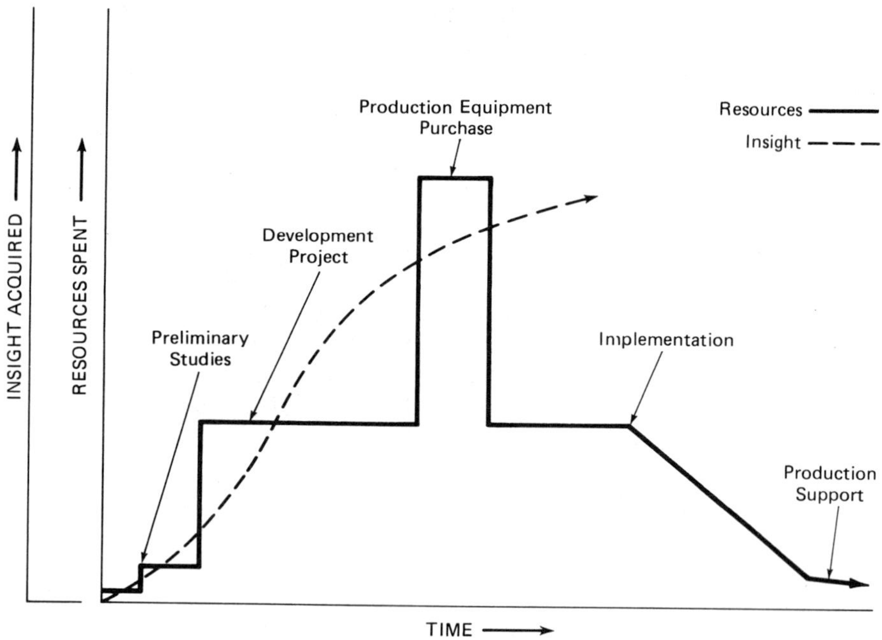

Figure 2.7. Resources and Insight vs. Time — Subcontract Applications Laboratory Approach.

development of the product and the technology for making it are impossible to separate—the manufacture of an integrated circuit, for example. But in many cases, the premature procurement of a particular type of equipment greatly increases the risks of project failure by locking the project into a given technology before it is really necessary, and by greatly increasing the time needed to conduct the project. If it is necessary to first complete the equipment procurement and setup cycle, then conduct the development project, and then repeat the procurement cycle to obtain a production facility, the amount of technically nonproductive paper shuffling is obviously doubled. With the current rate of technical improvement, where CNC and laser manufacturers, for example, are producing a new generation of machines every two to three years, this ponderous method of project management is just too slow to assure a company a sufficient competitive lead to justify the increased resources it requires.

In most cases, the first machine a company actually owns should be the one which is implemented in production to produce a profit. How this can be done, while actually reducing rather than increasing the chances of failure, is explained in the following chapters.

The common assumption that a manufacturing technology project can only be conducted if an expensive facility is first procured is based on the erroneous idea that successful manufacturing technology projects result from facilities. Success results from competent and dedicated people, and facilities are only the tools such people use.

Starting a project with an equipment procurement prevents the immediate investigation of its technical questions, the answering of which is essential to success. Instead, months or even years can be spent in the frustrating clerical work needed to get approval of capital appropriations to procure the equipment. This certainly does not help a project to gain the necessary momentum.

If this is the only acceptable method for conducting an advanced manufacturing technology development project within a company, it has one more important disadvantage; it provides a convenient excuse for those who wish to appear interested in advancing the company's technological position but do not actually want to conduct any projects, either out of laziness or fear of failure. They are free to talk about what great innovations they would produce if only they could get the equipment they need to start.

No one needs equipment to start a project. One needs first an idea, then a commitment to providing experimental evidence that the idea has merit. The equipment to do this can usually be borrowed or rented. Experimental evidence of the value of an idea and the suitability of the proposed equipment to the company's products should always be required before any substantial amount of facilities money is spent. But once the necessary experimental evidence is available to justify the next step, whether it is further development work or procurement of a production facility, the decision should be promptly made and approved. It should then be relentlessly executed.

Another common problem with manufacturing technology development labs is a tendency, especially in companies where they are strongly supported, to become isolated from manufacturing operations. In its extreme form this can lead to the finding of elegant and expensive answers for which no questions exist. The natural tendency of those primarily assigned to these laboratories is to spend all of their time in them and to see anything but the briefest and cleanest of visits to manufacturing as distractions. This tendency should be strongly discouraged. Manufacturing is the customer of the applications laboratory. Like any customer, his problems with the things previously sold him cannot be ignored if he is expected to buy the next product.

The question of whether or not to use the applications lab approach for a given new process development should not even be considered until a rational feasibility study has been conducted and the central technical questions to be answered by the program have been defined. It is only then that one can decide if this method should be followed or if some other, less expensive and less time-consuming technique should be used.

REFERENCES

1. Kidder, Tracy. *The Soul of a New Machine,* Boston: Little, Brown, 1981.
2. McMurray, J., et al., U.S. Patent 4,442,968 (April 13, 1984).
3. McMurray, J., et al., U.S. Patent 4,394,347 (July 19, 1983).
4. Tuchman, Barbara W. *The March to Folly, from Troy to Vietnam,* New York: Knopf, 1984.

3
PRELIMINARY EXPERIMENTS AND FEASIBILITY STUDIES

CENTRAL TECHNICAL QUESTIONS

The first step in doing anything successfully is to understand exactly what is to be accomplished. The next step is to define the obstacles to be overcome, and then decide what methods will be first tried to overcome them. Someone has an idea; he suggests that some new technology just leaving the laboratory might provide an improved or a completely new way of doing something. Or he suggests some novel combination of existing technologies.

A simple, relatively quick and inexpensive experiment to determine if the idea has any merit should be designed and conducted, with practically no fanfare and with no penalties if the idea does not work. This does not mean that one should rush out and try something without first assuring that the experiment is relevant to the questions involved. It must be planned intelligently.

Every step in the development process should begin the same way: with a one-sentence written statement of its purpose. In the preliminary experiment and feasibility study phases of a project, this sentence will probably begin, "To determine if it is possible to . . . ." In the full-scale development and implementation phases, this sentence should begin, "To develop and implement in production a . . . ."

Once the purpose is stated, the major obstacles to be overcome must be defined. These are the central technical questions (CTQs) that must be answered before the purpose can be accomplished.

The ability to perceive these CTQs and design simple experiments to begin to understand them is the most important skill in the early phases of a program. A good example of a device for defining and solving the CTQs of a particular project is shown in Figure 3.1, the Wright brothers' *Flyer*.

In order to accomplish their purpose, build a practical aircraft, the Wright brothers had to answer the questions of lift, propulsion, and control. The *Flyer* is a device brilliantly designed to answer those questions and no others. It has no wheels, because they already knew how to make wheels, so they wasted no weight or time or money on perfecting the

Figure 3.1. The Wright Flyer.

wheel. They also left until later the questions of galley design and in-flight movies. They did, however, define the questions of lift, propulsion, and control and find a set of working answers to those questions. Other, better answers were found later, but none were found until the questions were first defined and answered by the *Flyer*.

The CTQs of a development program are usually as obvious after success as they are to anyone now studying Wrights' first airplane. To be able to comprehend them at the beginning of the program or to at least perceive their existence and construct the early phases of the program to find and deal with them is absolutely essential. If one is not dealing with the CTQs of the project, one is not working on the project at all.

The CTQs can often be found by asking the question, if this does not work, what will be the most likely technical reasons for its failure? The goal of program management is the efficient conduct of programs to maximize the gains from success and to minimize the costs of failure. One of the best ways of maximizing the cost of failure is to postpone the tough and important questions while investing heavily in things that will have no value unless production-suitable answers are found to the CTQs.

Preliminary experiments and feasibility studies are the only method by which ideas should be selected for development. Selection based on expert opinions or committee decisions is worthless. Only experimental evidence of its suitability qualifies an idea for further development. Launching a development program, even one based on an excellent idea, without first defining the CTQs and constructing the project to answer them can produce very expensive failures. Even if they eventually succeed, such programs always take far longer and cost far more than necessary.

At this, the very beginning of the project, it is a mistake to be overly concerned with whether or not it will succeed in the form in which it is conceived. The most commonly asked questions are, how soon and how cheaply will it work, and what will be the return on investment? The question which must first be answered is, will it work, and if so, how? In the worst case, the innovator or man-tech manager is forced to guarantee that it will work before spending any money to determine if it will work. Until the preliminary investigation is complete, it is usually not possible to begin to assess the risk or the possible payback.

The frequency of program failure for technical reasons should decrease as programs progress, as is shown in Table 3.1. A well-run advanced manufacturing technology department will conduct many preliminary experiments, and not all of these will succeed. If none ever fails, something is very wrong. Either too few ideas are being generated and tried, or something is discouraging experiments which push at the front of technology. If every preliminary experiment succeeds, active steps should be taken to find the cause and correct it.

Fewer feasibility studies should fail, but occasionally, some should. But once a formal development project has begun and the building of a pilot plant begins, the chances of failing to produce and implement a production system should be quite low.

Table 3.1. Frequency and Acceptability of Failures of Well Run Manufacturing Technology Projects.

Project Phase	Failures Are:
Preliminary Experiment	Frequent and Completely Acceptable
Feasibility Study	Slightly less Frequent and Acceptable
Development Project	
• Pilot Plant	Uncommon but Tolerable (Replanning and Modification more Common than Failure)
• Production System	Extremely Uncommon and Unacceptable
Production Implementation	NEVER AND COMPLETELY UNACCEPTABLE

Once a project has built and operated a successful pilot plant, project failure for technical reasons is impossible. If a program reaches this point without production-suitable answers to all of the CTQs being found and verified, its failure is not for technical reasons at all but because of poor management.

Why would someone conduct a project while avoiding the central technical questions upon which its success ultimately depends? For excellent political reasons. In an organization where everything must always succeed, the pressure to avoid failure can be overwhelming. Avoiding the CTQs and doing the parts of the program that are not technical but clerical is one way to be able to show quick superficial progress. If management has an immature need for instant and constant success, and perceives the existence of problems as proof that those seeking answers to them somehow erred, only sham projects that develop nothing will be conducted. This situation sometimes results from the original idea being generated by an upper management which is so sold on it that they cannot tolerate a feasibility study which shows that perhaps some other approach might be better.

Avoidance of the CTQs can also occur when an idea is oversold to upper management by a middle manager before being proven. He may then evade the CTQ which he did not answer before announcing his breakthrough. If government or corporate funding is obtained without first conducting the experiments needed to assure that the proposed approach will work, no one is overly interested in discovering that there is a CTQ for which no answer exists.

The CTQs can be incorrectly defined and therefore incorrectly answered if the statement of purpose is too limited in its scope in the early phases of the program. For example, suppose one is faced with a high wear rate on a component which requires use of a very expensive abrasion-resistant material. A program to find a way to increase the wear resistance of less expensive materials might be proposed. If the statement of purpose were written by someone who wanted the job and was considered to be a laser expert, it might say, "Develop a laser system for heat treating the component to obtain better wear from a less expensive material." The statement, while it might be suitable for a full-scale development program after a proper preliminary study is conducted, avoids the central technical question which must be answered first; namely, what is the best way of hardening the material? It should say, "Evaluate possible methods of improving the abrasion resistance of less expensive materials and select the most promising technique(s) for further development." Obviously, the first could lead into a technical and political dead end from which it could be very difficult to escape if substantial funds had been spent.

Another way in which an incorrect statement of the purpose can lead to the CTQs being incorrectly defined or answered is by limiting the program to a single manufacturing operation and ignoring the interrelationship of that operation with the rest of the manufacturing process. An example would be developing a machine capable of processing 1000 parts per minute which is supplied by an existing operation which produces 10 parts per hour. If the statement says, "Develop a machine for such and such an operation . . ." instead of saying, "Develop a process for manufacturing such and such a part . . . ," this can happen.

In these cases, one aggressively attacks the wrong problem or the right problem in the wrong way. In some organizations, announcing to management that something has been done aggressively is all important, and management either doesn't care or is incapable of understanding if it made sense, as long as it was done aggressively. In such companies, someone who takes the time to design one or more rational preliminary experiments, and then uses the results to design and conduct a rigorous feasibility study, may appear to be indecisive.

The difference between logically working to obtain the information needed to plan a successful development and implementation program, and slowly strangling the idea by demanding test after test until the idea dies, is sometimes not obvious to those who do not have discernment about these matters. This is why it is important to have the specific test requirements and minimum limits for success stated before the tests are run, as part of the design of the experiment.

THE AGT 1500 TURBINE ENGINE RECUPERATOR

The AGT 1500 engine powers the U.S. Army's main battle tank. Its specific fuel efficiency is improved by the recuperative design of the engine (Figure 3.2).[1]

In turbine engines of conventional design, air is taken in at one end and passes through a compressor. The compressed air then enters a combustor where it is mixed with fuel and ignited. The resulting train of hot gases then passes through the gas producer turbines which drive the compressor, and then through the power turbines which are connected to an output power shaft in turboshaft engines or a fan module attached to the front of the engine in turbofan engines. The gases, with most but not all of their heat energy extracted, then are exhausted from the engine.

The AGT 1500 engine is a recuperative turboshaft engine. In such an engine the air from the compressor does not flow directly into the combustor; it first passes through a large heat exchanger—the recuperator. Here it is separated from the exhaust gases by only 0.008-inch-thick nickel-

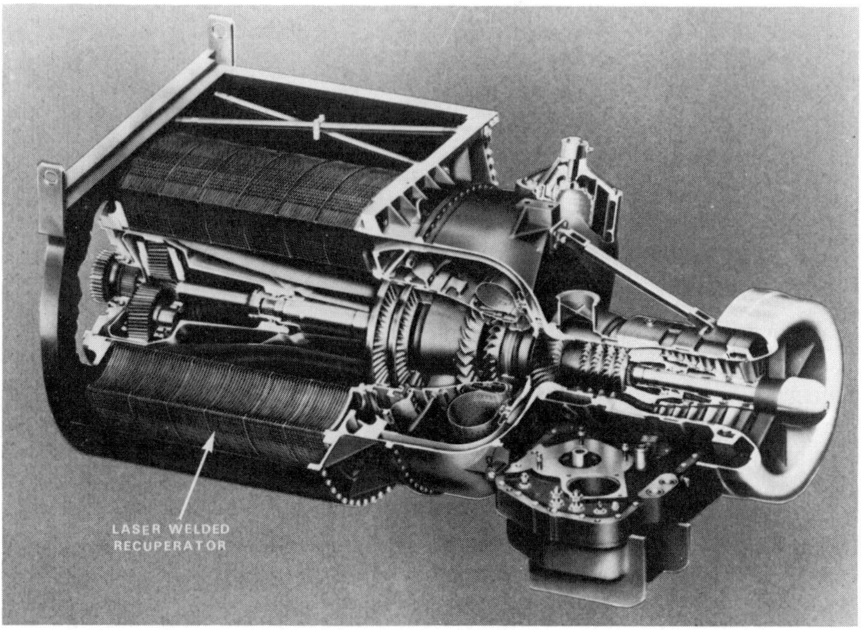

Figure 3.2. AGT 1500 Turbine Engine.

based steel. Its temperature is thus raised by extracting waste heat from the exhaust gases. After passing through the recuperator, and being preheated, the air flows to the combustor, where less fuel is thus required to heat it to the engine's operating temperature. The gas train then passes through the gas producer and power turbines and finally through the exhaust gas passages in the recuperator, where its residual heat is extracted before being exhausted overboard.

The recuperator is shown in Figure 3.3. It is an all primary surface anular heat exchanger designed to give the best possible heat transfer between the compressor air and the exhaust gases with the least pressure drop. It is 27 inches in diameter, with a 16-inch-diameter center hole. It is made of 600 separate plates of the 0.008-inch-thick material, 300 having an embossed waffle pattern (Figure 3.4) and 300 having a channel pattern (Figure 3.5). When these are stacked alternately and welded about the air passage holes and the inner and outer circumferences of the anulus, they form both the radial passages for the exhaust gases and the circumferential passages through which the compressor air flows.

The manufacturing process for the recuperator begins with coiled sheet stock being fed into a progressive die press or being sheared into squares

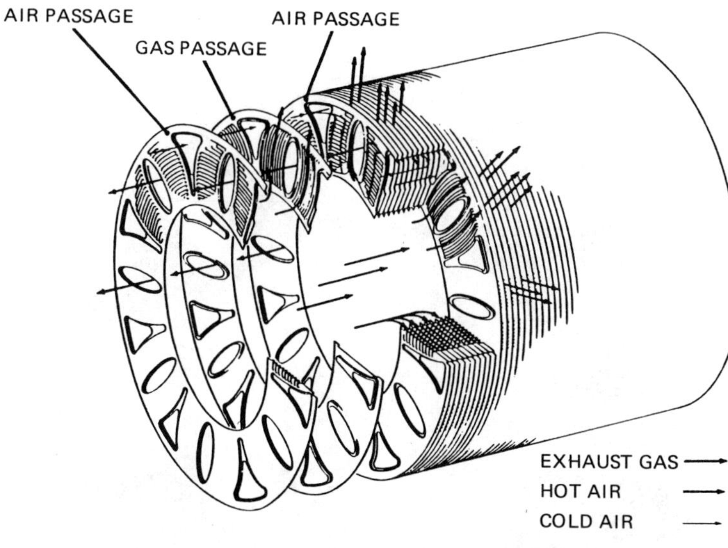

Figure 3.3. Recuperator Schematic.

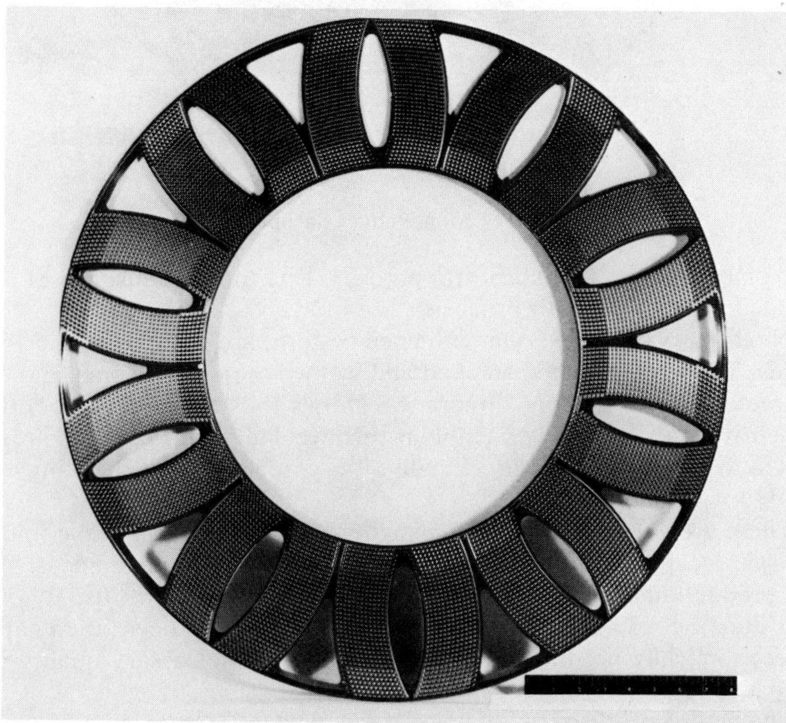

Figure 3.4. Waffle Pattern Recuperator Plate.

Figure 3.5. Channel Pattern Recuperator Plate.

and embossed by the hydroform process. Due to the thousands of plates needed per month, both processes are used.

Next, pairs of plates, one with each pattern, are welded around the air holes. The pairs are then stacked and welded around the inner and outer diameters of the anulus. Figure 3.6 shows these assembly operations schematically. The core assembly is then welded to a header and pressure tested to assure its integrity. Finally, it is assembled in its housing and is ready for engine installation.

There are about 6500 feet of gastight weld around the air hole joints of a single recuperator and about 3500 feet around the circumferences. When the production of the AGT 1500 engine began in the late 1970s, all joining was done by resistance welding. This was the only process then capable of economically producing miles of welds to the necessary quality standards.

The recuperator is probably one of the most difficult components in the world to make. For every recuperator core produced, 600 plates of a ma-

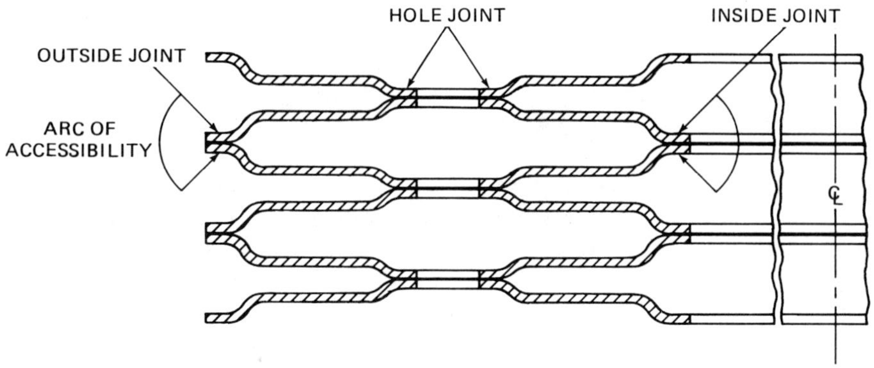

Figure 3.6. Recuperator Assembly and Welding Schematic.

terial that is very difficult to shear because of its thinness and ductility must be cut and formed and joined with over 1.5 miles of weld. Well over 100 cores per month are routinely manufactured.

When the recuperator was designed, resistance seam welding was the only practical method for producing those miles of weld. In this process, two copper electrodes connected across the secondary of a transformer squeeze together the sheets of metal to be welded. An electric current passes between the electrodes and through the metal. This produces heat at the area of highest electrical resistance, the interface to be joined, causing it to melt. When the current flow is stopped, the molten metal at the interface freezes, making a weld.

The original recuperator production facility used special automated resistance seam welding machines. This process has a number of inherent limitations which affect both productivity and quality. It has an upper limit of speed, and the contacting copper electrodes have poor wear characteristics. The weld quality is affected by the instantaneous condition of the electrodes, by the surface resistance of the materials to be welded, by the internal resistance of the heads which support the electrodes, and by the amount of squeeze pressure actually being applied.

The welds are very difficult to inspect because they are at the interface between the plates and can only be evaluated by cutting cross sections. This forces the use of inspection by assumption. A sample or a part is cut up, and if found acceptable, the parts made by a supposedly identical technique are assumed to be good but are not themselves inspected. A complicated quality control structure is necessary to assure that the bases of this assumption are true.

The recuperator and its manufacture have been explained in some detail because several of the advanced manufacturing technology programs that will be used as illustrations in this and the following chapters are concerned with it.

PRELIMINARY EXPERIMENTS

The purpose of a preliminary experiment is to determine if the proposed idea has sufficient merit to justify a more formal feasibility study. The experiment should be designed to determine quickly and cheaply if the solution suggested to the supposed central technical question has any value. If this seems vague, it must be, in a true development project. If all of the technical questions and their answers are known at this, the very beginning of the project, then it is not an advanced manufacturing technology project.

The welding of the inner and outer circumferential joints of the recuperator by laser welding was suggested. This was thought to be a major improvement over resistance welding, because delivering the welding energy to the joint as a beam of light rather than through the electrodes would resolve the inherent conflict in materials properties of resistance welding electrodes. The high electrical conductivity needed for resistance welding requires the use of high copper alloys, which have poor wear properties. If, however, the edges of the plates could be melted together with a laser beam, the wheels which clamp the plates during welding would not be electrodes and low electrical resistance would not be required. Wear-resistant materials could thus be used. This would increase welding speeds and reduce machine maintenance. It would also give a weld that was visible for inspection, thus reducing the complex and costly process controls needed to guard resistance welding's inspection assumptions.

The first question to be answered was, could the proposed weld be made between two pieces of the recuperator materials at a welding speed that would promise a reasonable improvement in production rate? The preliminary experiment was designed to find an answer to this question. Two strips of recuperator material were clamped together between two aluminum bars (Figure 3.7) and trial welds made at three times the speed possible with resistance welding. It was found that a laser was capable of producing such a weld.

This experiment was not designed to answer all of the central technical questions, only the most important: Could a laser produce the required weld under the most ideal conditions? During actual welding of the inside and outside joints of a recuperator at the speeds proposed, the problems of tooling to maintain contact between the joint faces and present them to the beam and the problems of real-world plate edge condition would have to be solved. But until it was known that a weld was possible in the

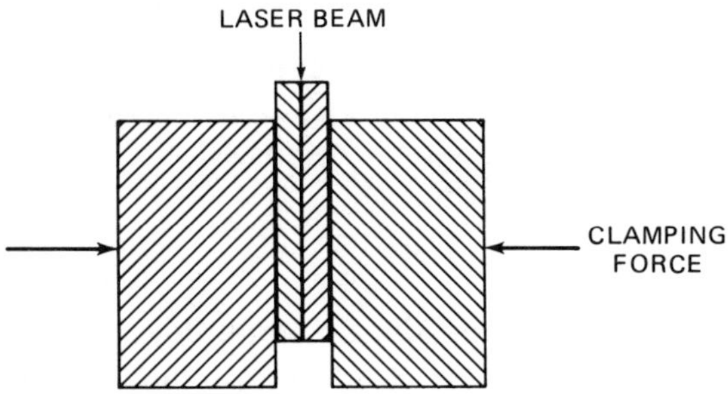

Figure 3.7. Recuperator Welding Preliminary Experiment.

ideal situation provided by the flat strips clamped between the bars, it was pointless to address these problems. The experiment was designed, and it succeeded in demonstrating that an answer to the most central question was possible. It was done quickly and cheaply.

It was not done carelessly. I was once involved in a program that proceeded from a similar preliminary experiment to full-scale development without a feasibility study. After months of being unable to duplicate the successful test weld made in the experiment on actual hardware, we found that the wrong material had been used in the experiment. The person who had conducted it had somehow substituted a much more readily weldable material. When the preliminary experiment was duplicated with the correct material, we found it to be a failure. By then a great deal of time and money had been wasted on something which was not even within reach of the technology.

Every bit as much care and professionalism must go into the planning of preliminary experiments and their conduct as would be used in a multimillion-dollar development project. This planning must clearly state the purpose, the materials, the procedure, and the minimum criteria for success. Quick and cheap is not sloppy.

It is quite common to have a failed preliminary experiment that does not end the project. Often the information gained from the failed experiment will suggest a different experiment which will succeed or which will fail and suggest yet another experiment. This is not the same as conducting one experiment which fails and then adjusting the test plan or doctoring the results to convert it into a success.

Sometimes the pressure for a success is overpowering. Very seldom does anyone look for a replacement for an operation or a process which runs perfectly. Man-tech programs are aimed at the areas where the biggest improvements are possible; those are the worst problem areas. The more difficult the problems with the existing process, the more the pressure for a success if it is known that a better technique is being investigated. Preliminary experiments are therefore best conducted very quietly. When the experiment has shown that the central questions may be solvable, planning of the more detailed feasibility study begins.

FEASIBILITY STUDIES

It is bad management practice to begin a full development program for an advanced manufacturing process without conducting a detailed preliminary study to determine if an implemented production system is technically possible and economically desirable. This preliminary project is required to provide the insight into the central technical questions of the program needed to plan, fund, and conduct the full-scale development program. Besides technical feasibility, the preliminary study must also investigate the economic viability of the proposed innovation. Although this study may not generate hard financial data to precisely estimate the payback, it should give some idea of its promise. These cost data will of necessity be approximate, but they should not be discarded or ignored. There is a tendency to reject estimates given in round figures, while numbers reached by even less rigorous methods are believed if they are stated to the penny. The study should be planned to provide data from which reasonable estimates can be made.

A policy of always conducting a feasibility study and basing the decision to conduct a full-scale development program upon it encourages the suggestion of ideas. People can be kept from making suggestions by too manic an acceptance as much as by rude rejection. If it is feared that management will launch an expensive, poorly planned program based only on one's suggestion and then hold one responsible for its failure, only those things which are absolutely sure to work will be suggested. That is not innovation.

Planning and Control of Feasibility Studies

An important criterion in conducting a program is that no one has done it before. If the results of an identical program, reliably conducted, can be obtained, obtain them. If it has never been done before, obviously uncertainty and risk are involved. This inherent uncertainty is what makes rigid planning and control bring failure despite obsessive attempts to avoid it.

If planning and control are too loose, in the hope of encouraging innovation, programs can wander into a state where the big breakthrough is always just one phase away. In such programs, long scholarly reports about obscure theoretical developments can become the only product.

These conflicting requirements for control and flexibility cannot be met by finding some moderate technique and applying it to all parts of the program. Each program element should be planned and controlled to a degree of detail and flexibility suitable to it.

The elements of an experiment are taught in elementary science courses: purpose, materials and apparatus, procedure, results, and conclusions. The purpose must be succinctly and specifically stated. No change should be permitted once the program is begun. If the purpose is changed, the original program is no longer being conducted. Perhaps, as the program progresses, it is discovered that the original statement of purpose is not technically possible nor is it desirable. This is not uncommon. One must then go back and review the preliminary experiments, plans, and assigned resources to determine if they are appropriate to the new purpose. If they are, they were probably inappropriate to the old one. The reason the purpose must be succinctly written is to prevent it from being subtly changed to fit the most recent set of results or to circumvent a difficult CTQ. Steady modification of the purpose is the key to conducting truly aimless projects. Here, one must be absolutely rigid.

The material upon which the new process is to operate must also be defined in detail. Whatever it is, metal or plastic or cloth or glass, one must assure that it is the same that will be expected to enter the process in the production shop. This is one area where it is absolutely necessary to know what truly takes place in the shop. No engineering drawing or specification should be believed until it has been verified by actual investigation. Do not depend on the assurances of manufacturing management that everything is what the paperwork says it should be. A production suitable system must operate on the production materials with all of their real flaws and variations. If specimens are used, be sure that they represent the worst case expected. Therefore, material requirements should have wide but rigidly set limits. Here the tendency is not to open the limits as the program progresses, but to narrow them, so that the more difficult to process materials are eliminated. The material section of the plan is thus another element that should be defined in detail, and modification of the requirements should not be permitted without reviewing the program to see if its justification and logic are still valid.

The apparatus to be used is a project element where great flexibility should be allowed. Allowing this flexibility is one of the main arguments for using competing equipment manufacturers' laboratories rather than only the equipment available in one's own applications laboratory. In these preliminary programs, where equipment costs are intentionally kept as low as possible, the cost of trying several apparatuses in parallel programs is minimal. The only firm rule in this element of a program is that the equipment concepts involved should be readily automatable and production suitable. Perhaps, in a part of the study, material is cut with a saw. This can be done with a hand saw in the feasibility study if it obviously can be replaced by a power saw in the production system. If, however, the manual device appears to be the only way to do it, stop right there. Another central technical question has been found and must be answered.

The procedure element of projects must be subdivided into two parts. The first is the procedure for developing the process, and the second is the procedure for evaluating the product of the process. Great flexibility is required in the first and great rigidity is required in the second. Because no one has before done exactly what is being attempted, the best way of doing it may not be obvious at the beginning. The fact that the most promising procedure will probably be modified and improved as the program progresses is no reason for not stating it, analyzing it, and trying to improve and simplify it on paper before committing resources. Flexibility and unplanned chaos are not the same thing. However, just because considerable work is put into a plan, it should not become immutable. Developing a well-thought-out plan without allowing it to become a straitjacket requires an honest appraisal of what is actually known, what is surmised, what is guessed, and what is only hoped for.

The planned development procedure should be written with the expectation that the least understood parts will be changed the most, as insight is acquired. As with the apparatus, if there appear to be two or more possible procedures, the use of parallel programs to simultaneously evaluate them should be considered.

The procedure for evaluating a new process and its product, and the minimum results for success, must be clearly stated prior to beginning the program. If they are not, no rational program exists. State all of the requirements for success: product quality and the methods of evaluating it, production rate and the methods of measuring it, unit cost and the method for calculating it, or any other characteristic that will be part of the decision to begin the development phase. Often these criteria will be based on a comparison with an existing technology, and the most common problem in this comparison is obtaining data not on the new but on the old. Major cost elements may have been hidden over the years in overhead or been traditionally transferred to other operations. The apparent quality and costs of existing operations may have been synthetically improved by hiding the scrap and rework costs associated with them. In some cases, these things have gone on for so long that they appear rational to everyone concerned. It is usually much easier to generate good data to evaluate the new technology because this data acquisition is a planned part of the project. One should never accept, without questioning, data on a running operation or process. Make acquiring, or at least checking, these data also a planned part of the project.

Once the criteria of success have been defined in the project plan, everyone who will later be involved in evaluating the program's results must agree that these are adequate, and that if they are fully met, no further testing will be required. Often there is pressure to oversell the program's promise in the beginning in order to get the approval of some less courageous manager for the feasibility study to begin. Do not oversell. Work out reasonable minimum criteria for success, and agree to them and exceed them. But do not let someone bid up the program so that, if absolutely everything goes perfectly, it will just barely succeed. This is a common technique to get the innovator into a no-win situation. It is often accompanied by arbitrary cost estimate cutting, so that more and more for less and less is promised in order to keep the idea alive. Kill the project right there, or better yet, complain that those using this technique are trying to strangle the idea before it is even tried. Then suggest that since they think a successful program can be conducted with these trivial resources, they should conduct it. When avoiding being bitten by a dog, it helps if one is at least a little smarter and a little meaner than the dog.

The conclusions element of a program is simply a statement that if the agreed criteria are met, a proposal for a development program will be prepared and the feasibility study will be considered a success. If they are not met, the program will either be terminated or another feasibility study, using a different approach, will be conducted.

Selection, Justification, and Funding of Feasibility Studies

Selection and justification of a manufacturing technology project is not a single step in the development process, but a continuing part of it. A recent study conducted by the National Electrical Manufacturers Association[2] determined the most common criteria used by companies in justifying manufacturing automation. It found that the most important criterion was financial payback. The next was improved product quality, followed by long-term benefits, operational safety, and environmental concerns. That these other factors—quality, safety, environmental concerns, and the long-term health of the company—are considered to be separate from financial payback shows that the methods for evaluating payback are rather crude. These things all have great impact on profit, but because they are not easily quantified, they are not included in its calculation. This exclusion is acceptable as long as it is understood that this is the result of deficiencies in accounting practice and not that these factors are dealt with less specifically because they are less important.

This problem is symptomatic of a more fundamental problem: the belief that those things which do not appear on a balance sheet do not exist and therefore do not contribute to the bottom line. Actually, when a financial report shows that a company sold so many of a given item in some period of time, it is really giving quantitative data on the perceived value of that product to the customer. Value is the quality of the design and the quality of manufacture of the product related to price. There may be a difference between the perceived value and the actual value of the product, but as customers become familiar with it, the difference will generally disappear, no matter how much advertizing money is spent. Thus the number sold, especially when compared to the performance of competitors, is a measurement of the quality of the product. Quality is thus a major factor in the bottom-line performance of the company, even if disguised by methods of accounting.

When, however, one must prepare a predicted financial report for something which has not yet been done, the problem of how to include critically important but usually unstated factors can be insurmountable. This is why the selection and justification of new manufacturing innovations strictly by the estimated bottom line of a return-on-investment calculation is a mistake, unless the effect of improved quality on sales or the value of the technical insight gained can somehow be included in this calculation.

There are numerous accounting techniques for producing detailed estimates and using these to evaluate the suggested project's potential economic value to the company. In adapting these techniques for advanced technology projects, it must be remembered that the final conclusion, no matter how precisely stated, is no more exact than the least precise guess made in the data used to produce it.

If the estimate of risk, for example, is based on subjective perceptions, it is probably better to state them in prose, describing the logic behind them, than to hide their inherent imprecision behind a pseudoscientific probability calculation ultimately based on the same subjective perceptions.

This is not to say that no financial estimates can be prepared or should be required, but that the natural limitations and inherent difficulties should be accepted and resolved. While it is acceptable (barely) for a manager to base a decision about past performance purely on the bottom line and ignore the factors which comprise it, it is idiotic to use the same technique when using estimates to decide about something that will or may happen in the future. The manufacturing process change which is most readily justified by a strict return-on-investment calculation is the one which should not be selected, the project which reduces production cost by increasing productivity while ignoring quality: the production line speedup. Production rate is the easiest thing to predict and to increase, as long as quality is ignored. Yet quality is a major factor in maintaining and increasing market share. Thus, each time a project is evaluated, not only its estimated cost and financial return, but its effect on quality, worker and environmental safety, and the long-term benefits of the knowledge gained in conducting the program must also be assessed.

The overall concept of man-tech program management upon which this book is based has built into it a rigorous system of program selection and evaluation. At the end of the preliminary experiment, a decision, based upon criteria decided before the experiment was conducted, is made whether to discard the idea, proceed with a more detailed feasibility study, or repeat the preliminary experiment in some different way. At the end of the feasibility study, the program is again evaluated. As explained in later chapters, decision points are also planned into the development program. As insight is gained, more detailed justification is required. Preliminary experiments require essentially no justification. If they succeed, their results are used to justify a feasibility study.

This justification should still be relatively simple and only semiquantitative. If, after a simple experiment, someone can concoct detailed cost figures to justify a large investment into a full-scale development project or system procurement, either it is a reinvention or he is a snake oil salesman. At the end of a preliminary experiment, no matter how successful, no rational person can or would want to justify a full-scale program. Nor should he have to.

A well planned and conducted feasibility study should be relatively inexpensive and quick. Assuming full-scale programs costing (including the first production equipment) from $1 million to $2 million, preliminary studies might cost from 1 to 10 percent of the total, with most costing less than 3 percent.

The justification for the feasibility study should consist of a detailed explanation of the preliminary experiment and its results. It should then explain the way that this innovation will affect the product and manufacturing process. Next, the proposed study should be explained, following the format given above (purpose, material and apparatus, procedure, results, conclusions), with emphasis on the presumed central technical questions and how they will be approached. The estimated costs of the study should then be explained, and the total amount requested for the study stated. Complicated projected cost studies should not be required at this point, because in a well conceived and managed program, the data needed for them are not yet available.

The entire program proposal should not be much more than two typewritten pages. If it is necessary to write long background statements about the existing manufacturing process or the product to be made, the wrong people are doing the evaluation. Those doing the evaluation should be very few and very familiar with the background.

The proposal should only be evaluated by those capable of understanding it. Manufacturing and marketing people should eventually become involved in any project which will have an effect on them. However, at this stage, where so little is certain, it is too early to require their approval before proceeding. They can be asked for their comments, but if one forces them to approve, they will probably want to know as much about the new process or product as they think they know about the old. Confusion about the difference between approving a feasibility study to investigate a new process or product and approving the process or product itself is quite common and surprisingly difficult to dispel.

The accounting department should never be allowed to approve of or select these projects. In the better opera houses the ticket takers are not allowed to select the program or conduct the orchestra.

A company which wishes to encourage manufacturing innovation, yet assure that its resources are not wasted on pie-in-the-sky programs, must carefully choose those who will select and approve these studies. Their job will be made at least possible if they are presented with the information in the proposal as described above. But to evaluate it, they need discernment about the people making the proposal as much as about the proposal itself.

For these reasons the program approval process should require that preliminary experiments be approved by only one person, very close to where ideas are generated: the lab and the shop floor. The approval of feasibility studies should involve this person, but should also require justification to someone not directly involved, yet having the necessary background understanding, so that he can independently evaluate the information. This person is probably his boss or department manager.

Once the study is under way, prompt feedback of cost data is essential, because if the stated upper cost limit is reached, a rejustification of the project should be required. It is neither automatically canceled nor allowed to go on spending money based on vague promises to do better. The project, if the spending limit is reached and the agreed criteria for success are not met, must be completely reassessed.

Reassessment should be done when about 90 percent of the amount of money originally thought to be enough has been spent without the program's being completed. This reassessment should be conducted in a rational, professional manner. It is very poor leadership practice to become accusatory with someone who has been attempting to invent something because his original estimate of the resources required was too low. This reappraisal, if properly conducted, can save a foundering program. If conducted improperly, it can help to kill innovation in a company.

LASER WELDING OF RECUPERATOR CIRCUMFERENCES—FEASIBILITY STUDY

The purpose of this study was to determine if laser edge welds with high-temperature fatigue strength equal to or better than resistance seam welds could be consistently made joining two AGT 1500 recuperator plates. The approach used was to weld internal pressure test specimens which simulated the welding conditions and service loading of the recuperator plates (Figure 3.8). Considerable baseline data on both resistance seam welds and laser seam welds were available from an earlier successful program to develop an automated laser welding facility for the air hole periphery joints. These specimens, while not nearly as large as recuperator plates, required over 18 inches of weld each, and thus, in the quantities needed for fatigue testing, provided a reasonable weld sample size.

During testing, specimens are restrained between two stainless steel plates in the same way that a single pair of plates is restrained in the recuperator. They are heated to the operating service temperature, and the interior of the specimen is alternately pressurized with nitrogen and vented. Test pressure and cycles to failure are recorded and analyzed using standard statistical techniques. The failure areas and unfailed areas of all tested specimens are metallographically evaluated.

Figure 3.8. Recuperator Internal Pressure Specimen (Pillow Specimen) with Smooth Edge Weld.

Welding of the specimens was conducted in the applications laboratories of two laser manufacturers. Both were paid for their efforts, but all welding was done with presence and contributions of the engineer who had originally suggested the program. Two vendors were used in order to avoid becoming wedded too early to a single technical approach. The customer's engineer was present because he knew the most about the idea and understood the requirements of the recuperator manufacturing system. This insight was absolutely necessary to the conduct of the study.

The welding setups used by both vendors were originally similar. The test specimens were clamped between aluminum holding dies and rotated under a vertical carbon dioxide laser beam.

The first welds did not go well. One of the ground rules was that the edges of the plates were not to be perfect but to have the burrs and mismatch that would be expected in production recuperator plates. This was an added requirement over the preliminary experiment, which used carefully prepared and aligned edges. When these more realistic joints were hit with the laser, very nonuniform energy coupling took place. Many tries with various combinations of parameters yielded very poor results. The best that could be done was a saw-toothed weld (Figure 3.9) made by grossly overpowering the weld and ignoring the reflected energy. This technique would not have been suitable as a basis for developing a production system. If no better way had been found, the program would have

Figure 3.9. Pillow Specimen with Saw Tooth Edge.

been a failure—an inexpensive failure, though, because no equipment other than the simple holding dies had been built.

The program was not a failure, however. The customer's engineer had considerable experience in electron beam welding and had once been involved in using that equipment to weld materials similar to those in this program. Because it is readily deflected by electromagnetic fields, it is common practice to deflect the electron beam so it describes a small high-frequency circle while moving along the joint. This makes the welding of thin materials with mismatched edges possible by spreading the beam power over the edges of both plates rather than concentrating it at the interface.

The customer and vendor engineers built an erector set rig for deflecting the laser beam in a small circle. Because the laser is not deflected by electromagnetic fields, the concept used to deflect the beam, a rotating lens, was different. But it worked. The laser now consistently produced uniform welds between nonuniform edges. The beam energy coupling problem was solved.

A number of specimens were welded with this technique and returned to the customer's plant, where they were tested. The results were not good. These welds, in spite of their attractive appearance, had a fatigue strength of only about 20 percent that of resistance welds in the same material. Metallographic examination showed the failures to be through the weld bead. The welds made with the other techniques routinely failed through the base metal (Figure 3.10).

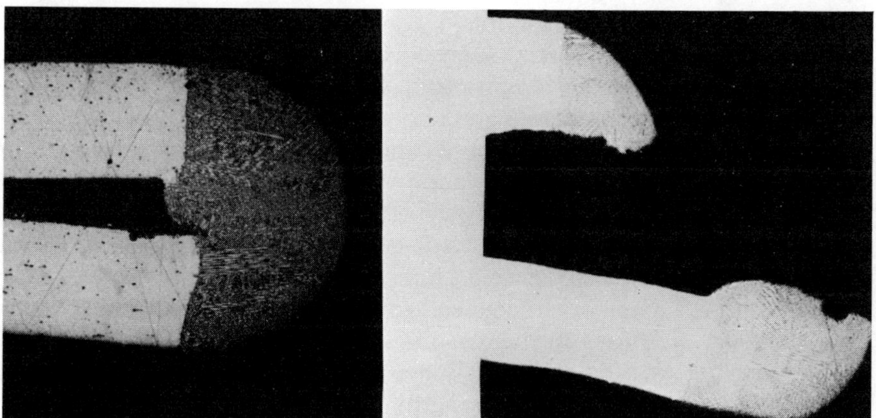

(a) As welded (b) Fracture

Figure 3.10. Weld Made with Deflected Vertical Laser Technique.

Perhaps the program would still fail; these welds were obviously weaker than welds of about the same size made by the other methods. However, the problem had been transformed by the development of the circular deflection technique from a question of how to make a weld to how to make a strong enough weld. Like an onion, problems are peeled away one layer at a time until nothing remains.

The program was being conducted by the man-tech department manager who was also the program engineer because the original idea had been his. The testing and evaluation of the specimens was being conducted by the most junior engineer in the department under the supervision of the supervisor of welding development. Supervision is not exactly the right word; with the assistance of, or with suggestions from, would also describe the relationship.

This situation, the department manager being the program engineer and the most junior engineer conducting the evaluation can, in an organization where failure is not tolerated, be ethically unhealthy. In such groups the scientific method is defined as determining what results the boss wants and then designing the test so that only those results are possible.

Because the test conditions and the minimum results for success were rigidly defined before testing started, this was not the case. The unwritten rules of the group stated that anyone showing such intellectual dishonesty would not be tolerated, and the rules about freedom of speech assured that at the first sign he would be made aware of it. This is extremely important, because without these norms of behavior, it is impossible to have the aggressive program pushing which is essential to success, without the risk of elastic professional integrity. In this case, the junior engineer, although she showed the requisite concern, quietly enjoyed announcing to her boss's boss that his onion problem had yet another layer.

All the information available on the project was assembled, and a meeting of all the engineers in the department was called to try to find an answer or to end the project if none could be found. At such meetings any idea can be attacked, but no person.

Because a number of those attending were not directly involved with the project, a detailed explanation had to be given by those conducting it. This, in itself, can sometimes lead to the finding of a possible answer to a technical question. In this case, the central technical question was that the edge welds would obviously need substantially more cross section in order to have the required strength, but the circular beam deflection technique seemed to limit the size of the weld possible. Yet, the circle was necessary in order to make a weld at all.

The meeting developed a possible answer: a two-pass weld. First, the circular beam would be used to join the plates and produce a uniform surface; the second pass would be with a straight beam which should couple with the now homogenized surface and produce a substantially larger and stronger weld. Such a technique would be possible in a production machine by using two laser beams, a short distance apart, with the plate edges first passing through the rotating beam.

This worked. Specimens were welded with the two-pass method with the parameters of the second pass set to produce a much larger weld (Figure 3.11). Again the junior engineer conducted the fatigue tests, and again she was able to announce that the onion contained yet another layer. Although the fatigue strength had increased to 65 percent of the minimum requirement from the 20 percent of the single-pass weld, it was still well below the strength of the resistance welds which was the minimum value for success.

Again all of the information was reviewed, and again the review provided the answer. Every tested specimen had microsections made and examined through the area of failure and through the areas which survived the test. Photographs had been made of all of these, similar to those in Figures 3.10 and 3.11. These were correlated with the fatigue strength of each specimen. The mean fatigue strength is statistically derived from the results of a substantial number of similarly made specimens. When the individual specimens having the higher strengths in the population were compared with those having lower strengths, a cracklike surface between the unmelted inner face of the plate and the drop of melted metal which formed

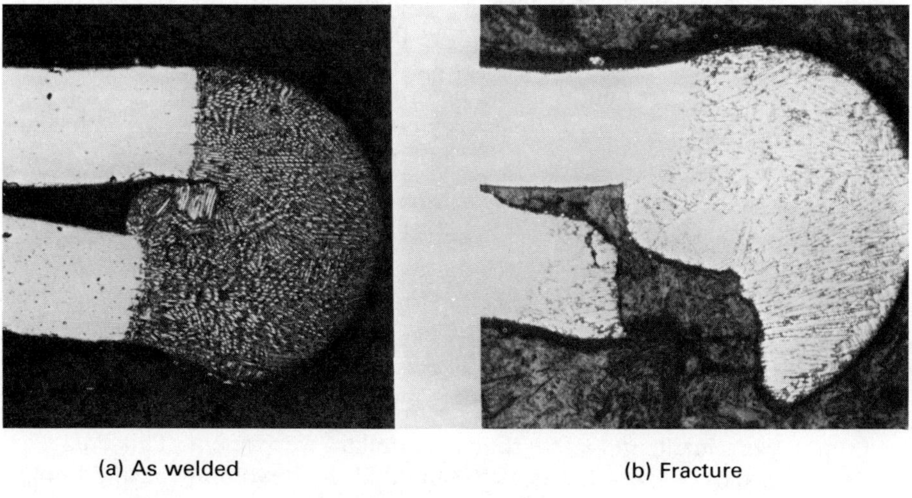

(a) As welded

(b) Fracture

Figure 3.11. Weld Made with Two Pass Vertical Laser Technique.

between the plates during welding was noticed to be more pronounced in the weaker welds.

Welding of the specimens was being done on a rig with the laser beam vertical because this is the standard beam orientation in laser welders. The drop of metal which flowed downward between the plates during welding caused a cracklike cold shut which appeared to be the reason for premature failure. This might not happen if the welding beam were horizontal. The actual recuperator plates would have to be welded in the horizontal position because of their size, weight, and flimsiness. Therefore, the rig was modified so that the specimens could be welded with horizontal laser beams to find if this would change the weld solidification pattern and eliminate the cause of the premature failures.

The simple laboratory setup had grown with each layer peeled from the onion, and now it seemed that most of the C-clamps in the Bay Area were involved in the project. But the required number of specimens were made with the horizontal beams and returned for testing. These yielded results consistently in the upper portion of the acceptable scatter band and had a mean fatigue strength 110 percent that of resistance welds. The feasibility study had succeeded.

THE CONDUCT OF FEASIBILITY STUDIES

This program succeeded in dealing with the CTQs of the process because it was conducted in a way that assured they would be addressed. The criteria for success and the test conditions were rigidly defined; yet the development procedure was flexible enough to permit the program to follow the path of technical logic as that path was discovered. This is not easy, and the logic of the project is a great deal more obvious now than it was to the project engineer just before the idea of circularly deflecting the beam in order to solve the energy coupling problem occurred to him or when the low strength of the resultant welds was first reported. Intelligent management of these projects does not state in advance the route to be followed. It does, however, rigidly define the goals and the rules to be followed in reaching them.

The statement of purpose of this project was "to determine if laser edge welds with high-temperature fatigue strength equal to resistance welds can consistently be made in the material of the AGT 1500 recuperator."

In the preliminary experiment, a few hundred dollars were spent and no equipment was bought. In the feasibility study, some simple holding fixtures were made and a few thousand dollars were spent. This cost was about one percent of what it was estimated the development and implementation of the production system would cost.

This program was based on specimens rather than actual parts. There is always risk involved in using specimens, because specimens which are substantially different from the part in question may very well mask CTQs. This program was not the first venture into welding recuperator plates with a laser. A totally automated system for welding the air hole joint peripheries (see Chapter 4) had been developed and successfully implemented. That experience provided the insight which minimized the risk of performing the feasibility study on pillow specimens alone. The feasibility study of that earlier program had been conducted on both pillows and sections of actual plates because many of the central technical questions of that program involved finding ways to deflect the beam around the irregular-shaped holes at high speed. An excellent rule is to always try the idea on actual hardware as soon as possible. In this case, since the CTQs of welding a part could be simulated with the specimens, and the cost of building tooling to manipulate the actual part was so high when compared to the tooling for the specimens, it was decided to take the risk. However, the first task in the development projects was to build the necessary equipment for testing the techniques developed in the feasibility study on full-size recuperator plates.

The central technical question of the study was expected to be the fatigue strength of the welds. Several authorities had given excellent reasons why a weld at the edge of the plates could never be as strong as one further in and of the older configuration. The energy coupling problem was not expected, because the preliminary experiment had given no hint of its existence. This does not mean, however, that no problems were expected.

The objective of the program and the rigidly defined criteria for success were based on high-temperature fatigue strength, because if that were insufficient, the most elegant, efficient welding system in the world would be useless. Once the joint strength question had been positively answered, it was known that the new system, with its inherently more inspectable weld, would make not only a cheaper recuperator but a better one.

The quality of the new product or the product of any new system must be at least as good as that of the old. In most cases, the new should be better. A more efficient way to make a poorer product is not an improvement, it is merely a way to pass the company's unanswered CTQs on to its customers faster. There are few things in this world so vengeful as a customer sold an unanswered central technical question.

Two vendors were originally contracted, in this study, to do the same task in parallel. Whenever the basic technology is rapidly developing, as are lasers, one should always start with parallel development. In this particular case, one vendor gained an early lead by his ability to quickly devise a setup for oscillating the beam. The other vendor did not produce such a system, choosing instead to attempt to find other ways of solving the coupling problem. These were not successful. The money spent on giving this second vendor an opportunity was not wasted, however. Insight into his capabilities and equipment was gained, and understanding of the things which were unsuccessfully tried are also of value. None of this would have been gained if the customer's engineer had not been working in the vendors' labs with their people.

There is a tendency, when using suppliers' applications laboratories for preliminary experiments and feasibility studies, to award a contract and allow the vendor to do the work by himself. This is very common when the work is being done at no cost. It is bad practice. Even the best vendor, and the two involved in this study were both excellent, is motivated by profit, not the dispassionate search for scientific truth. A basic rule of these studies is that the methods used must be suitable for development into a production system. Vendors have been known to use obscure methods which violate this rule in order to land an order. The potential customer's people must be there to assure that this does not happen. Occasionally, one will encounter a vendor who will not allow the potential customer to see the work being done, usually for reasons of trade secrecy. Offer to sign a nondisclosure agreement; if he still refuses, go elsewhere.

This program was an excellent example of the value of getting people with diverse skills involved. The laser manufacturer's people knew a great deal about lasers and their application. The customer's people knew a great deal about the manufacturing processes. All of the skills available were needed to solve the problems. One is buying knowledge in these programs. If the customer does not have his people where the knowledge is being generated, in the lab, what is being paid for will not be acquired.

The first use of that knowledge, after the completion of a successful feasibility study, is the planning and preparing of the proposal for the development project. It is obvious what a powerful proposal can be generated by a feasibility study such as the one just described. A potential funding source is generally much more impressed by a solid understanding of the key problems and demonstrated solutions to them than by vague assurances that nothing can go wrong.

REFERENCES

1. Miller, J. A., and Chevalier, J. Development of an automated laser beam welding facility for high volume production. *Welding Journal,* **62**(7):49–54 (July 1983).
2. Brosilow, R. Coming soon to the production floor . . . . *Welding Design and Fabrication,* 38 (June 1984).

4
PRINCIPLES OF MANUFACTURING TECHNOLOGY DEVELOPMENT PROJECT MANAGEMENT

INTRODUCTION

It is surprising how often major programs to develop improved manufacturing techniques will be planned, funded, and conducted without consideration of the best way to do the job. These programs are not only expensive and have great impact on the long-term survival of the company, but they are subject to the law of lost opportunities. This law states that when limited resources are spent on something which fails, not only the resources are lost, but the opportunity to conduct another, successful program, and all of the advantages of that success, are lost as well. The actual loss from conducting the wrong project, or the right project badly, is therefore many times the cost of the project.

Questions such as when and if prototype systems should be built, how interim goals should be set to facilitate both flexibility and control, and how to divide systems into subsystems and projects into phases so that interface problems are reduced and not aggravated must be rationally answered. Incorrect and sometimes patently hopeless approaches to projects often have a superficial logic which makes them appear, to the inexperienced, not so much as the best way to do the job but as the only and the obvious way to do it. Any given project can be planned and conducted in many different ways, and no two projects are exactly the same. Yet, there are certain general principles and techniques derived from them which, when followed, improve a program's efficiency and reduce the chances of failure.

Fortunately, although it did not seem so at the time, I was once involved in a project which illustrates these principles as though it were designed to do only that.

RECUPERATOR LASER WELDER—PHASE I

The fully automated laser welding system for the recuperator air hole periphery joints was one of the more difficult and most successful man-tech programs done in recent years.[1,2] Three complete systems are now running in production, welding over 30,000 joints per day, totaling over five miles in length. Phase I of the program was unsuccessful. Although it was not then apparent, the technical problems encountered were the result of basic mistakes in the management of the program.

The original design of the recuperator was developed using 6-inch-diameter models, simulating one set of air and gas passages and a single air inlet and outlet hole. These models were electron beam welded together in a laboratory in the late 1960s. Electron beam welding, although it produced excellent welds of recuperator material in a laboratory, was not a viable production technique. It requires high vacuum, and the problems involved in automating the loading, unloading, and manipulation of the recuperator plates in a vacuum chamber were insurmountable. From all of the processes surveyed, resistance welding was selected as the only one capable of economically making the miles of weld required.

These resistance welds were made using an automated system in which seven separate welding machines were arranged around a large carousel which indexed the recuperator plates from machine to machine. Each performed one of the welding operations required to join a hole pair (see Figures 3.4, 3.5, and 3.6): ellipse left side, ellipse right side, triangle outer edge, etc. The basic concept led to a machine of great mechanical complexity which was difficult to maintain (Figure 4.1). Because its index function was based on the dimensional accuracy of the entire recuperator plate, it was intolerant of tolerance accumulation in the press dies which made the plates. This made the process difficult to control when using plates, which, of necessity, came from several die sets.

In the middle 1970s, lasers began to appear that were supposedly capable of operating in a high duty cycle production environment. They seemed to have the advantages of electron beam (EB) systems: high welding speed, low heat input, no need for contracting electrodes, without the disadvantages that made EB welding unsuitable for the application. Therefore, a preliminary investigation was conducted.

An authority on lasers was consulted who announced that a 2-kilowatt constant wave laser was the most suitable for the application. Such a laser was available at a research institution, and it was used to produce six internal pressure specimens (Figure 3.8). When these were tested successfully, a welding system was ordered.

This, of course, was the first mistake. This preliminary experiment did not concern itself with the central technical questions, which were, even then, obvious. These were the ability of the laser to operate dependably at the required duty cycle, the problem of guiding the beam at high speed

Figure 4.1. Recuperator Hole Joint Resistance Welder.

around the irregularly shaped holes, and the problems of part manipulation and tooling.

This program approach can be called the procure-then-develop, or the buy-then-try method. It is the turnkey procurement technique with a development step inserted between the delivery of the equipment and its production implementation (Figure 4.2).

The system ordered was not the complete apparatus to weld recuperator plates. It included the laser, the moving mirror system which deflects the beam around the joints, and the computer and software which controls the mirror motion and welding functions. This system supplied everything required down to the beam output nozzle of the system. This approach was based on experience in earlier programs and with other manufacturers of such equipment and their usually limited knowledge of turbine engine materials and welding applications.

This technique of project management had worked in the past because both the customer and the equipment manufacturer were dealing with the part of the system each understood best. Unfortunately, in this case neither party understood nearly as much as they thought they did.

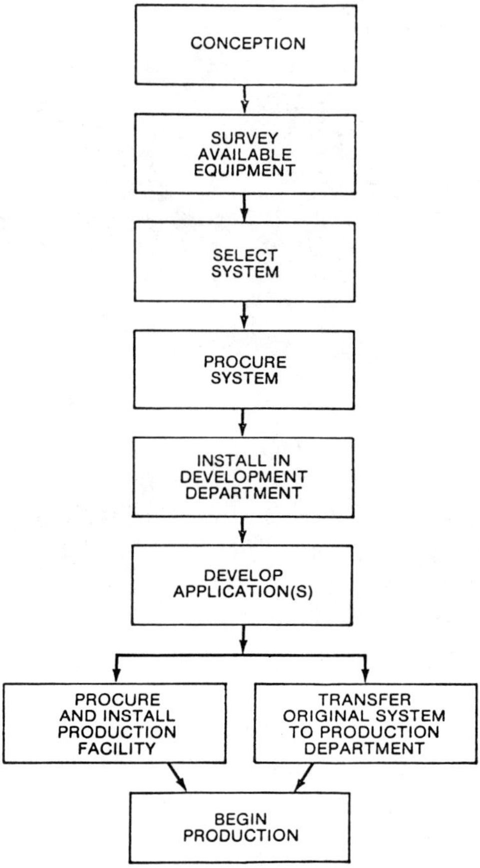

Figure 4.2. Procure-Then-Develop Project Management Technique.

In order to produce the required hermetic welds joining the plate pairs, the beam must travel at a high constant speed precisely around the rounded triangle and elliptically shaped air ducts. The periphery of each hole is a 0.080-inch-wide land of which about 0.040 inch is occupied by the gripping tools which hold the plates in contact during welding. The beam must traverse along the center of the remaining 0.040-inch path without drifting into the tools or the plate convolution. If it hits the tooling, a skip in the weld will result. If it hits the convolution in the plate, the plate will be cut because sufficient beam power to weld two plates will cut the top place if the two are not in contact.

The welding speed specified for the system was 235 inches per minute, 100 millimeters per second. This speed was chosen because the selected laser had successfully welded the specified material at about 80 percent of its maximum rated output at this speed. Although this was a rational decision based on the capacity of the laser, it failed completely to take the capacity of the computer numerical control (CNC) into account.

This device controlled the welding process and manipulated the beam by controlling the position, direction, and velocity of two axis tables to which the beam deflection mirrors were attached. Each axis was driven by a motor equipped with an optical encoder, which provided position feedback to the CNC, and a tachometer, which provided velocity feedback. Using these position and velocity data, the CNC would start, control, and stop the motion and turn the laser on and off as required to automatically weld the hole pair. It was only when the equipment vendor attempted to write the computer program that he began to comprehend the effect the high required velocity had on the computer.

The computer was selected by the equipment vendor. It used a 16-bit word and was programmed in a European numerical process control language. Allowing him to select both the computer and this language was also a mistake, in retrospect. But at the time it seemed reasonable, because he was responsible for and was warranteeing the system's performance. Input and output were through a paper tape reader and a mechanical teleprinter. (The system was ordered in 1977.)

By following instructions in the program, the computer generated signals to move the drive motors around the required path. Then, based on feedback signals, it issued corrections. Ideally, this process should take place as the system runs, in real time. Whether this is possible depends on the velocity of the mirrors, the frequency at which corrections are required, and the speed at which the central processing unit (CPU) and arithmatic logic unit (ALU) of the computer operate. The system was specified to operate at 235 inches per minute with a maximum track error of 0.004 inch. This meant that the smallest arc could be approximated by a straight line 0.027 inch long. The maximum processing time available is the length of one segment divided by the speed, 6.97 milliseconds.

The computer needed 350 milliseconds to multiply two 16-bit words. Division and the square root function were even slower. Because of these hardware constraints, real time processing was not possible and a complicated method of precomputing ideal velocity and displacement data was used. Figure 4.3 is a functional block diagram of the system.

A command table was generated by the input controller and used by the output controller as its program. The welding process was controlled by comparing the motor feedback signals to the precomputed ideal value. In this way, it compensated for the dynamic errors in the motor drive and mirror tables.

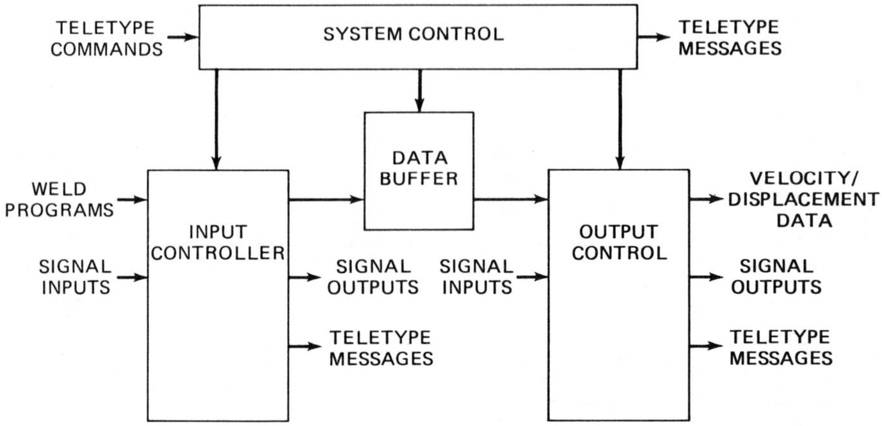

Figure 4.3. Computer Functional Block Diagram.

The first tests of the system showed a minimum error band of 0.024 inch, three times the acceptable value. The worst errors occurred on the upper right corner of the triangle joint and the ends of the elliptical joint. Here the path describes a small-radius turn requiring rapid acceleration on one axis and rapid deceleration on the other. Thus, the effect of a velocity change was not sensed before the next error compensation was required. This problem, too, was solved by making the computer program even more complex. A way was found to vary the amount of compensation generated, based on the increase in error since the last correction, by pulsing the velocity changes and by taking into account the response time of the drive motors.

The system was provided with an abort sequence which would stop it if a preset track error value was exceeded. When the system was repeatedly cycled with this program to verify its reliability, aborts occurred on about 3 percent of the runs. This meant that if it were in production, operating at the required rate, it would be down about 50 times per eight-hour shift.

Delivery of the system was now nine months behind schedule and not yet in sight. The project manager was under increasing pressure from management and the agency which had funded the program to somehow produce tangible evidence of progress. He visited the vendor's shop, witnessed an unsuccessful test of the system, and ordered it to be shipped nonetheless.

The principal investigator, who reported to the project manager, had had very little to do with the system, never having actually seen it. He had been mainly concerned with preparing the documentation needed to import it. The manager now replaced him as principal investigator with a more senior engineer who had previously accomplished several difficult production implementations of new technology.

The new principal investigator reviewed the status of the program and immediately suggested that a six-month extension be sought. When the system arrived, the project would have only two months left to run. This was not enough time to accomplish what remained. It would be impossible to do the tooling and welding development, which had originally been scheduled for the nine months lost on the vendor's computer problems, in two months. The request was refused because the vendor's lack of a sense of urgency was seen as the main cause of the lost time and the intent was now to keep the pressure on him. This was also the stated reason for ordering the machine shipped; once it was in the customer's shop, the new principal investigator, who was noted for his sense of urgency, was to set the pace for the vendor.

When the machine arrived, it was found that the vendor had more problems than an overly complex computer program and a deficient sense of urgency. The laser itself was unstable. It was necessary to constantly adjust the mirrors and output window in order to maintain a set output power. Eventually, this instability was found to be related to its basic design. The main power supply was found to be unreliable, as were the computer input/output devices. The computer program problems remained. The vendor spent another nine months flying people and parts back and forth from Europe, honestly trying to fulfill his purchase order commitment and warrantee. It proved to be impossible. Eventually, the defective parts of the system were shipped back to Europe for the major redesign and retrofit which this costly experience had shown to be necessary. The six-month program extension originally sought and refused grew to a year.

The system was made to work after a fashion. But it was never production suitable. Eventually, a fair settlement was negotiated to close out the purchase order. The vendor was scrupulously honest throughout the affair and did everything possible to meet his obligation. He just did not have the technical capability.

In any failure, there is plenty of blame to go around. Some years later it was learned that at the time the laser expert was telling the customer that only one system existed which would meet his needs, he was also trying to become the U.S. distributor for that system. The vendor was so anxious to introduce his system in the United States that he signed up for a level of performance impossible for the system to meet.

The project manager, too, made a number of errors. He depended on the expert opinion and did not investigate other potential sources. After tolerating the vendor's failure to deliver a working system until it was nine months overdue, he aggravated the technical problems for nontechnical reasons, by allowing the vendor to ship the unfinished system. If the vendor, in his own plant, with all of his resources at hand, could not make the system work, adding a transatlantic logistics problem was not going to help. The vendor's operating people went along with the decision to ship because they, too, were under management pressure to show positive results. Shipping the system was an important milestone, and reaching it became so overwhelmingly important that everyone became willing to suspend his intelligence in order to reach it.

None of these mistakes, however, was the basic reason for the failure. It was caused, first, by not conducting a proper feasibility study which addressed the central technical question of manipulating the beam around the small-radius corners at constant high velocity. Such a program would have been much more valuable than any expert opinion. It would have shown that the velocity at which the laser could weld the plates and the calculating speed of the computer were irreconcilable.

The second basic reason for the failure was that the procure-then-develop technique was unsuited for this program. The inclusion of the development step in the turnkey method is an admission that the technology is not completely understood and that development work is required to make it production suitable. This technique is suitable for projects where the equipment is well understood, inexpensive, quickly delivered, and, most importantly, not germane to any of the unanswered technical questions the development project is to answer. Projects such as new applications of truly well understood technology can be conducted in this way, if the engineering staff know exactly what is required of the purchased system, know how to specify it, and know from whom to buy it—really know, not just think they know. This allows the equipment to be ordered early in the program and the tooling or other system components to be built while waiting for delivery. The development step then consists of assembling the complete system, debugging it, and qualifying it and its product.

This management technique was completely unsuitable for this program, because none of its requirements were met. They were not even understood. A number of previous successful programs had been conducted in exactly this way, so it was adapted without even considering if some other approach might be more suitable. The pace of technical change and improvement has increased substantially in recent years. This has increased delivery times and the cost and complexity of equipment, while reducing the amount of available insight into these new processes. Thus both the risk and the cost of failure have grown. Yet, the procure-then-develop technique neither considers the risks nor provides a way to minimize and control them.

Technical risk is neither eliminated nor controlled in an advanced technology program by a performance guarantee from the supplier, no matter how legally airtight. If a cellophane rocket ship is ordered, no one is going to Mars no matter how brilliant the corporate lawyer. In this case, a warrantee was obtained, and the vendor did everything possible, at great expense, to fulfill it; but it was found to be technically impossible. Even if the customer is able to recover his purchase price from such a failure to meet a technical commitment, this is usually a small part of the resources invested in a project by the time it is finally declared hopeless and liquidated. Also lost are the competitive lead time and alternate opportunities which were not exploited because of the effort put into the failed project. These can never be recovered. Of course, warrantees should be required in any contract, if only to assess the vendor's true confidence in his ability to fulfill his promises. But a warrantee assures a program's success no more than a life insurance policy assures a long life.

The following is the actual statement of work proposed, approved, and followed in phase I.

1.0 *Process Screening*
1.1 Select a laser welding technique for more economical manufacture of recuperator assembly
1.2 Define assembly/welding method and the degree of automation required
1.3 Perform a preliminary cost analysis of the selected manufacturing approach

2.0 *Development of Optimized Technique*
2.1 Design and procure welding station and tooling
2.2 Set up and check out system
2.3 Conduct feasibility study by welding and testing representative subscale samples utilizing several candidate techniques
2.4 Select the best technique, scale up, and weld samples of full-size recuperator plates
2.5 Verify the selected technique through destructive and non-destructive examination.

The selection step (1.1) actually consisted of talking to the laser expert and running a few weld samples, but no actual parts, on the recommended laser. The definition of the assembly/welding method (1.2) and the cost analysis (1.3) were both paper studies with no experimental evidence upon which to base them.

Task two began with the design and procurement of the system. It was only after it had been set up and checked out that a feasibility study was to be conducted with subscale samples. The technique was then to be scaled up to full-sized parts.

It is easy now to read this statement of work and see its errors, just as it is easy to look at a 1902 automobile and see its design flaws. If the system had worked, this path could have led to a fully implemented production system. The problem was that it was predicated on nothing going wrong. If the expert really knew what he was talking about and gave an honest recommendation, and if those who designed the system knew exactly how to do it, then this program plan would have led to success.

But if all of this information was available, why was a development program being done? The title of the program was, after all, "Development of Improved Recuperator Manufacturing Techniques." This illustrates one of the crucial mistakes made in advanced manufacturing technology projects: failure to understand the difference between development and procurement. This program, as planned, was not a development project at all, it was to be an exercise in parts catalogue engineering. It was planned to buy the parts of the system, hook them together, and then conduct some tests to prove that it worked.

The problem here was that, while individual components were supposedly available to make up the system, no one understood the essential relationships between them. Without this understanding, a computer was chosen which did not have the computing power to control the process at the welding speed of the laser. The tooling was supposed to be designed and built without any insight into the constraints the actual laser welding process would cause. All of the crucial decisions had to be made before the information needed to make them could be generated.

Many man-tech programs never recover from the situation which faced the laser recuperator welder program at the end of phase I. The stated goals of the program had been to procure a system, develop welding parameters, and develop prototype tooling on which would be based the automated work handling system to be built in phase II. On paper all of these things had been accomplished. A machine had been procured. It worked poorly and parts of it were in the United States, while other parts were on a ship being returned to the vendor. But it had been procured. Welding parameters which produced acceptable welds had been developed, but they had only been run on flat strips of material by moving it under a stationary beam. The computer/moving mirror system was still incapable of actually welding a recuperator plate, and it was therefore unknown if plate pairs joined by these parameters would actually be acceptable. The development of the tooling had barely been started because the machine had never really been available for this work.

These things all resulted from the way in which the tasks were defined in the statement of work. The work was divided into tasks which had no specific technical results to define their successful completion "2.1—Design and procure welding station and tooling," really only required that some paperwork be processed and some hardware shipped. That the hardware be suitable or even work was not a requirement. The only task in the phase I plan which requires the generation of tangible technical results is the final one: "2.5—Verify the selected technique through destructive and nondestructive examination." Control of programs can only be maintained by dividing them into tasks which have specific technical results as their goals, so that interim technical progress can be evaluated. Placing a purchase order or loading some equipment on a truck is not technical progress, and attempting to use such things to gauge technical progress is nonsensical. The system would never have been shipped if task 2.1 had said "Design and build system and perform the following tests to assure the compatability of its components and its suitability for the intended purpose." The technical problems of the program were allowed to waste over a year because the management plan did not provide a means of recognizing them earlier and force either their solution or the program's redirection or cancellation. This should have been required not at the end of task 2.5, but during task 2.1.

The question now was what to do with the project? It could have been quietly allowed to expire by never beginning phase II, or phase II could have been begun as a continuation of phase I, going on with the hopeless effort to make the original system production suitable. The only advantage this option had was that it would eliminate the necessity of admitting the errors made in phase I. A major personnel reassignment occurred just then, and the principal investigator took over the duties of program manager. The very first thing he did, when informed of the change, was to begin looking for alternate laser equipment vendors and plan a study to evaluate their capabilities based on experimental evidence.

RECUPERATOR LASER WELDER—PHASE II

It was while reviewing this project and trying to find a way to salvage it that many of the interactions between technology and the management of technology discussed in this book were first noticed. It was decided to start again using what had been learned, not only about computers and lasers, but about how a project as technically advanced as this one should be managed.

The basic idea of laser welding the recuperator was reexamined and found to be as attractive as it had been at the beginning of phase I. It was obvious that the existing laser/computer system, although it was finally operating in the vendor's shop, would never be able to operate to the production duty cycle in the production shop. Another laser/computer system must be obtained. This time there would no dependence on the opinion of an alleged expert. First, the manufacturing process would be developed and thoroughly tested. Then the production equipment would be built and implemented (Figure 4.4).

This was a development project, and it would be conducted as such. The goal of a procurement is to acquire something as expeditiously and

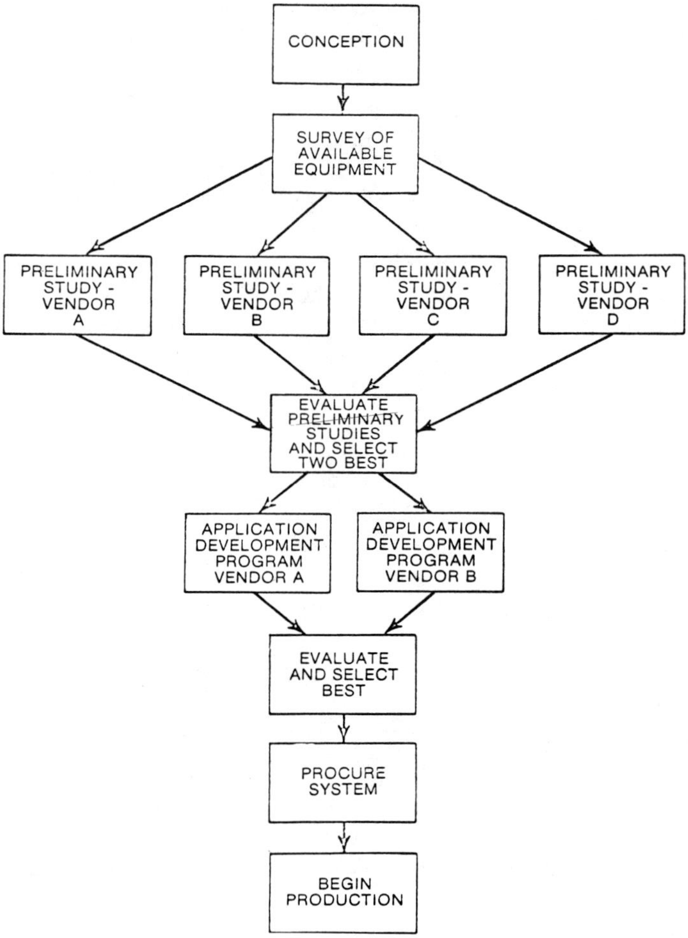

Figure 4.4. Develop-Then-Procure Project Management Technique.

cheaply as possible. Acquisition was secondary in the planning of the new project. If, and when, a system had been developed and proven to be production suitable, it would be acquired. But first it must be developed.

This development started by contacting every possible vendor of such systems in the United States. Three companies, in addition to the vendor from phase I, were found to be potential sources of such a system. All were interviewed, and their views on how such a system should be built were discussed. The solid knowledge of how not to build such a system, gained in phase I, was very valuable in asking the right questions. These meetings with potential vendors were conducted by the project manager. The purchasing department was informed but not involved because, as yet, nothing was being bought.

The four vendors seen to be potentially able to develop a suitable system were each awarded a small purchase order to build a simple holding tool to make sample welds of the elliptical hole joint. They used this tool and segments of actual recuperator plates to make sample welds in their applications laboratories. Figure 4.5 is such a tool.

This work demonstrated their ability to solve the central technical question of the project. It also provided the vendors with the understanding of the project needed to prepare two proposals. The first outlined costs, schedules, and techniques to be used in a program to develop a prototype

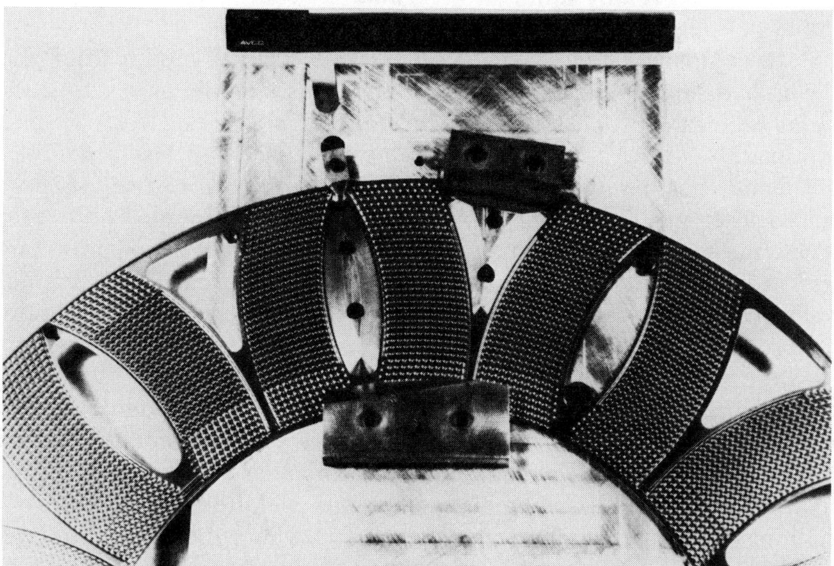

Figure 4.5. First Breadboard Laser Welding Tool.

laboratory welding sytem and use it to weld 350 pairs of plates. The second was a budgetary proposal for a production facility based on what was known at that time.

One potential vendor concluded as the result of this preliminary study that it was not feasible for him to participate and did not submit proposals. Two potential vendors were chosen from the three remaining candidates, and each received a purchase order to build the prototype setup and with it weld 350 plate pairs. This facility would be built in the vendors' applications labs and was not being bought. The laser, computer and moving mirror system would remain his. It was not intended that this sytem be production suitable, because as yet no one knew enough to design such a system. The simple setup (Figure 4.5) used in the preliminary study was referred to as the breadboard. The system, now to be built, was called the pilot plant. Only after it was successfully completed would ordering a production facility be considered.

It was understood that these pilot plants did not have to be totally automated as would be required of the production facility. However, it was required that only obviously automatable functions did not have to be automated. For example, the loading and unloading of the plates into and out of the prototype machine would be done manually, because this presented no technical difficulties. The actual welding cycle, however, including beam tracking, beam on and off, and beam to fixture location was required to be totally automated so that all central technical questions would be found and answered (Table 4.1).

Both companies now started to work: the original vendor from phase I, who had solved most of his earlier problems, and the new source. The technology each proposed to use was quite different. The original vendor continued with the concept of using a single 2000-watt laser, welding at 235 inches per minute. The second vendor proposed to use a 525-watt pulsed laser welding at only about 100 inches per minute. The required production rate of the production facility would be met by using two such lasers. The budgetary proposal showed such an approach to be economically feasible. In addition, the slower welding speed greatly simplified the problems of computer control by permitting real time processing of position corrections.

When each vendor had built, programmed, and debugged his setup, the first plate pairs welded were destructively tested to evaluate the weld quality. Next, 25 pairs were welded and then assembled using the production recuperator equipment to stack and weld them together around the inner and outer circumferences. This is where the final, unsolvable problem with the 2000-watt laser welding facility was found. This system produced so much distortion in the individual plate pairs that it was impossible to weld them together into a finished recuperator assembly.

Table 4.1. Automation of Functions—Production vs. Prototype Facilities.

Function	PRODUCTION FACILITY Device	PRODUCTION FACILITY Closed Loop?	PROTOTYPE FACILITY Device	PROTOTYPE FACILITY Closed Loop?
1 Load A Plate	Load/Unload Arm	yes	Manual	—
2 Load B Plate	Load/Unload Arm	yes	Manual	—
3 Position Plate Pair on Hold Fixture	Holding Fixture	yes	Manual	—
4 Close Holding Fixture	Holding Fixture	yes	Manual	—
5 Start Gas Purge	Gas Solenoid	no	Manual	—
6 Check Mirror Home Position	Moving Mirror System	yes	Moving Mirror System	yes
7 Start Travel Along Weld Path	Moving Mirror System	yes	Moving Mirror System	yes
8 Turn on Laser	Laser	yes	Laser	no
9 Control Beam Direction/Velocity	Moving Mirror System	yes	Moving Mirror System	yes
10 Turn Off Beam	Laser	yes	Laser	no
11 End Weld Travel	Moving Mirror System	yes	Moving Mirror System	yes
12 Index to 2nd Shape	Moving Mirror System	yes	Moving Mirror System	yes
13 Start Travel Along Weld Path	Moving Mirror System	yes	Moving Mirror System	yes
14 Turn On Laser	Laser	yes	Laser	no
15 Control Beam Direction/Velocity	Moving Mirror System	yes	Moving Mirror System	yes
16 Turn Off Beam	Laser	yes	Laser	no
17 Turn Off Gas Purge	Gas Solenoid	no	Manual	—
18 Open Holding Fixture	Holding Fixture	yes	Manual	—
19 Index Table to Next Hole Pair	Index Table	yes	Manual	—
20 Repeat Steps 3 through 20 Nine Times	—	—	—	—
21 Extract Completed Plate from Hold Fixture and Place on Finished Stack	Load/Unload Arm	yes	Manual	—

The other vendor, using the pulsed laser approach, produced pairs which were successfully assembled into a pack which passed all tests (Figure 4.6). The pilot plant then produced the 600 plate pairs needed to manufacture two full cores for engine testing. These tests, too, were successful. Only then did procurement of a production facility begin.

The building of the production facility was also done so as to gain actual operating experience as soon as possible. Although the basic layout of the welding stations was the same as the pilot plant, there were substantial improvements which came from experience with it. Rather than build the entire system, and only when it was completed start it and begin debugging, the key components of one of the system's two welding work stations were built first and installed on a laser in the vendor's applications laboratory.

This setup was used to weld about 2500 plate pairs while the rest of the system was being designed and built. A number of problems which could have complicated the debugging of the finished system were, in this way, found early and eliminated from the final facility design.

It should also be noted that, while phase I was a year late being completed, ran substantially over budget, and was a technical failure, phase II was completed on time and budget and was a technical success. Figure 4.7 shows the first production system. Figure 4.8 shows the welding tools and work station in operation.

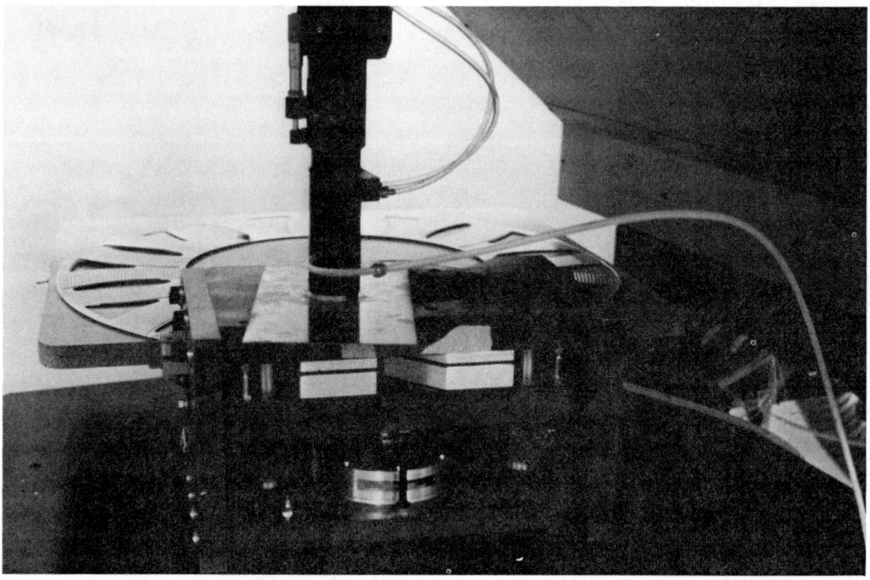

Figure 4.6. Pilot Plant Recuperator Plate Welder.

Figure 4.7. Production Recuperator Plate Welder Number One.

Figure 4.8. Production Facility Work Station and Welding Tools.

PRINCIPLES OF DEVELOPMENT PROJECT MANAGEMENT

The technique used in phase II worked. This does not mean that it should be used in this exact form for every man-tech program, or that every project manager is relieved of the need to think through his project and find the most suitable way to conduct it. There are, however, certain principles that can be derived, which have general application.

The first of these is that one must consider the management of the project as well as the technology to be used in the project to assure that both are suitable for the intended application. The management of technology is more than setting up budgets and preparing PERT charts. It is the intelligent commitment of resources based on a technical plan which assures that technical success is reached in the most efficient manner, with the minimum risk of technical failure. It seeks to minimize loss if technical success proves to be impossible. Although PERT and GANNT and networking techniques are useful devices, they are aids to intelligent planning based on discernment about the technology and its management requirements; they are not substitutes for it. Too often the technical planning of the project is separate from the financial planning and evaluation. Frequently, there is an adversary relationship between the two. The resources assigned to a project must be tailored to its technical requirements; the program must not be distorted to fit resources arbitrarily assigned to it.

A budget item where this arbitrary reduction of resources is common is in travel expenses for programs where major tasks are being conducted by subcontractors. In projects, such as the recuperator welder, that are conducted in the applications laboratories of subcontractors, the customer company saves considerable money through the elimination of the costs of operating and equipping its own laboratory for the project. These overhead costs are not completely eliminated, however. They are replaced by the travel expenses needed for the project staff to conduct the project at the vendor's laboratory. These essential costs must be included in the project budget.

Phase I was, to a substantial extent, planned with reduction of travel expenses as a major consideration. The vendor was located in the United Kingdom, and the project was allowed to drag on instead of confronting its problems because transatlantic telephone calls and telegrams were used as a poor substitute for on-site, hands-on control.

Phase II was conducted in California, and not only the project manager, but everyone who was expected to be involved with the system's implementation, visited the vendor's facility while the project was being conducted. These visits were not the usual quick review of the project against the plan, followed by lunch. They were of substantial duration, some several weeks long, and involved intimate involvement in the design and operation of the pilot plant, and later the full system. Although the vendor was responsible for the design and building of the system, the customer was responsible for its production suitability and the overall conduct of the program. Responsibility for complex technical endeavors cannot be discharged on a telephone.

To jeopardize a program by failing at the beginning to obtain the resources necessary for it to succeed is one of the worst mistakes a project manager can make.

The next principle of man-tech program management is concerned with how programs are divided into tasks and phases, and between the various organizations involved. Often, the most obvious places to divide responsibility and to separate tasks are the worst places to do so. Figure 4.9 shows this clean division of phases and those responsible for them in a project. It is a common, and usually fatal, mistake. Projects, like structures, fail most often at the joints. Good structural design minimizes the number of joints and locates them in areas of low stress. When multiple joints occur, they are not aligned with each other. These rules apply in exactly the same way to the design of projects.

There are two kinds of program joints: those between tasks or project phases, and those between various organizations involved in the project. These, if possible, should never be aligned as they are in Figure 4.10. For example, if separate departments of a company conduct the development phase and oversee the start-up of a new production line, both a task interface and an organizational interface occur together. If the organizational joint cannot be eliminated by having one department responsible throughout the project, it must be moved into a lower stressed area away from the joint between phases. This could be done by having the production implementation department enter the project during the development phase

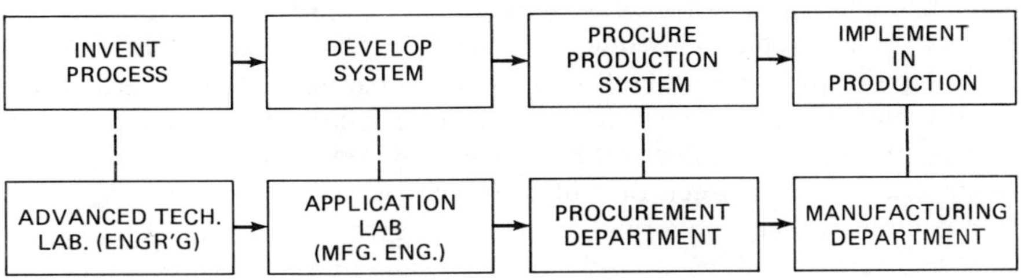

Figure 4.9. Neat Alignment of Phases and Responsibility Interfaces in Poorly Planned Projects.

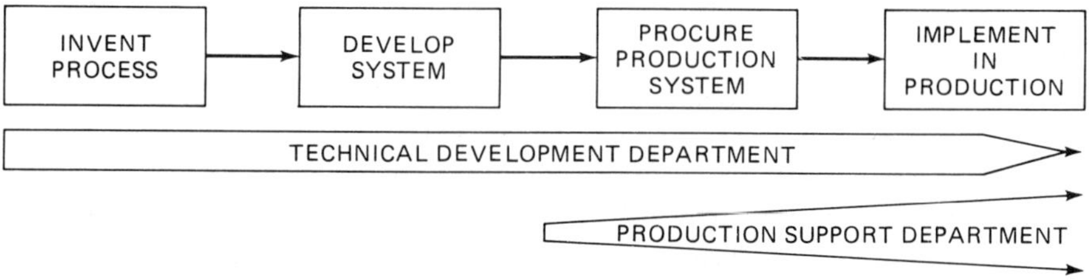

Figure 4.10. Offsetting and Lapping of Phase and Responsibility Interfaces in Well Planned Projects.

or by having the development department retain responsibility into the implementation phase.

The worst form of the interface alignment problem is to divide tasks and responsibility for them at an actual physical interface in an integrated manufacturing system. There is an almost hypnotic attraction to this neat and total division of the technical tasks, management responsibilities, and physical system along the same interface. It must be avoided. Whenever there is an unavoidable interface between parts of the project or parts of a system, the responsibility for solving the interface problems must not be divided. A single organizational unit must be responsible for both tasks or both subsystems and must develop the discernment necessary to integrate them flawlessly.

In phase I, the system responsibility was divided at the beam output of the moving mirror system. The laser manufacturer was to build everything on one side of that line—the laser, its power supply, the other accessory systems, and the moving mirrors—and integrate these with the computer. The customer would then develop the recuperator plate handling and welding tools and integrate them with the system. Figure 4.11 is the flowchart for the computer program for the finished production system. It is given here to show that attempting to separate the welding and tooling functions in an integrated computer controlled system was the same as teaching someone to walk by assigning the responsibility for each leg to a separate instructor. This does not mean that the separate sections of the computer program are not divided into separate tasks and dealt with in turn. But the tasks must all be the responsibility of the same people, because the subtle and difficult parts are in the relationships between the subprograms and between the software and the hardware.

In deciding upon an overall scheme for a given project and upon how tasks and responsibilities are to be divided, an important consideration is whether the project is to develop and implement a single manufacturing operation, as did the recuperator project, or whether it is to develop a

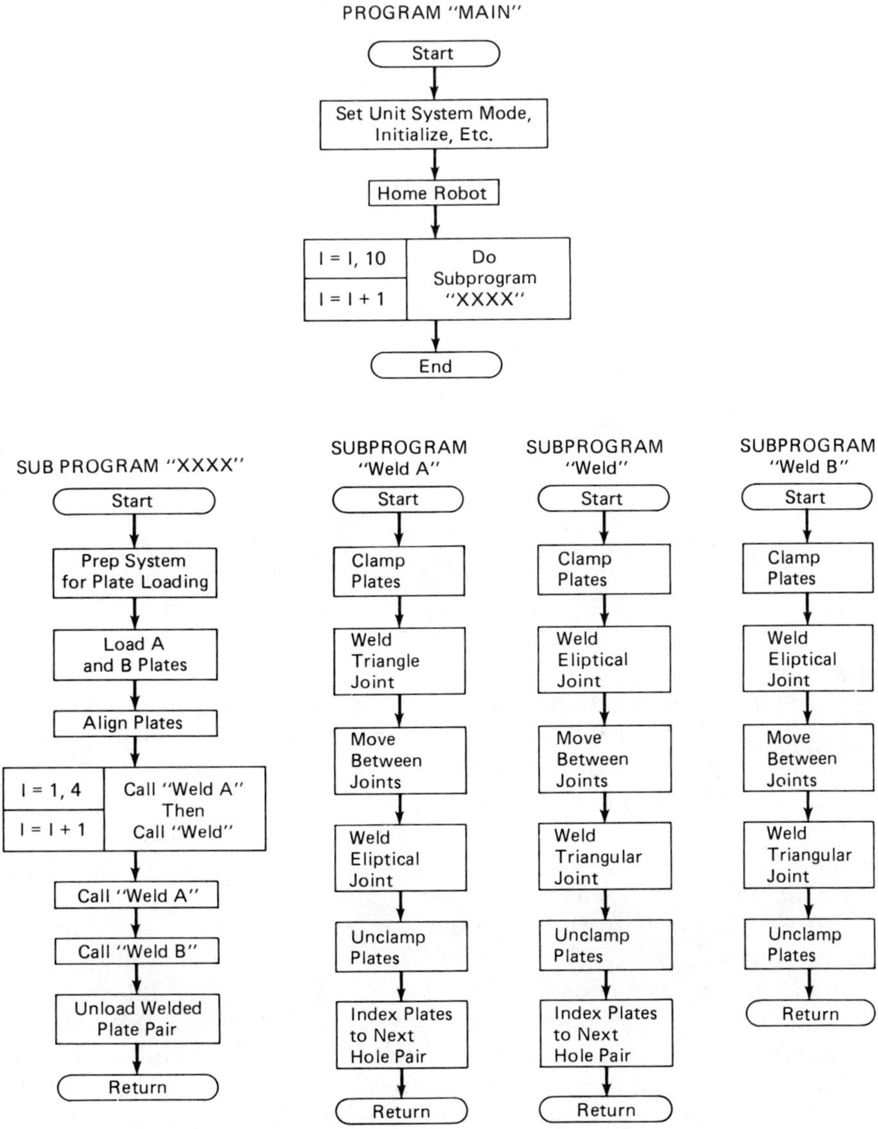

Figure 4.11. Computer Program Flow Chart — Production Recuperator Welder.

new process composed of a number of discrete operations for an existing product or component. The third and most complex case is the development of a complete process for an entirely new product.

In the second and third cases, it is necessary to divide the process into separate operations, and these may require very different skills and development techniques. In the latter two cases, even more than the first, failure will usually occur at the joints between areas of responsibility. Certainly it is here that most of the problems will occur. The design of the overall program must take this into account and assure that someone with the authority to solve these problems is in a position to acquire the insight needed to recognize and solve them.

An important technical necessity in the planning of advanced manufacturing technology projects is to build a prototype or a pilot plant or a working model which deals with the area about which the least is known, just as soon as possible. There is no substitute for the knowledge that can only be gained by actually trying to make the process work. If, in phase I, this had been done, all of the serious problems with the original system would have been found much earlier, at much less cost. That it was not until the building of the prototype system at the beginning of phase II that we discovered that the original system was incapable of making components which could be made into acceptable assemblies proves that some sort of prototype should have been built early in phase I, before any system was ordered.

Admitting the need and planning for an early prototype is an honest admission of ignorance and is sometimes unacceptable for personal or organizational reasons. Although it may present a decisive image to management to claim to know everything needed to be known before the project starts, in the long run it is much better career management to admit ignorance and build a prototype to dispel it than to be a short-term budget meeting hero by eliminating its cost.

If the project is to succeed, and if it is truly aimed at developing an advanced technology, its basic concepts must be tested as soon as possible and as rigorously as possible. The project plan must be built around this need, and the project funding must provide for it. In phase II, we actually built four simple preliminary prototypes, one by each vendor, then built two complete prototype systems before actually deciding on the design of the production facility. To do less would have been to base phase II upon the same thing upon which phase I had been based: an ill-informed guess with no experimental basis.

The progressive building of systems, starting with the primitive breadboard for the feasibility study and ending with the implemented production system, is how advanced manufacturing technology development projects must be constructed. Knowledge is gained from each version, and this knowledge is used to make a decision to proceed or to abandon the project. If it proceeds, the next version is based upon the previous one. Although the systems become more complex and complete as the project progresses, each is easier to build than its predecessor because the most difficult problems are dealt with first in the earliest prototypes, and the last parts added to build the production-suitable system are the least challenging technically. For example, the prototype system built by the successful vendor of phase II had a completely functional laser, moving mirror system, and welding tooling, all controlled by a computer (Figure 4.6). It welded almost 1000 plate pairs before the decision to build the production system was made. The production system consisted of these key elements with the addition of a pick and place load/unload device and, with some previously manual functions, such as shielding gas on/off, automated (Figures 4.7 and 4.8).

However, this system had to be production suitable, which was not a requirement for the prototype. Production suitability means that the system is designed to be operated and maintained, while dependably meeting a production schedule, by production personnel. This requirement is probably the most difficult for any advanced manufacturing system to meet, and designing such a system without first having all of the technical questions solved and demonstrated by a working prototype is generally impossible.

When a complex system is developed, a prototype will be built. If no prototype is planned in the project, the alleged production system will be that prototype. Solving the central technical questions usually involves piling modification upon modification. This makes it impossible to even consider the factors necessary to production suitability while building the prototype. These factors include accessibility of components for repair, selection of components for maximum dependability and life, and simplicity and safety of operation. Trying to solve these problems in the design before the answers to the CTQs are completely understood is generally not possible. Those systems which finally worked in a laboratory, but are unused a year after entering production, are often not so much failures as successful prototypes from which no production system was ever built.

There have been some recent impressive advances made in the computer modeling of manufacturing systems by robot manufacturers. Although such models help to solve problems of equipment layout and location, and allow the basic positional sequence programming of the robot to be demonstrated on a cathode ray tube rather than with built up systems, they are not the equivalent of a working prototype. For example, no computer model can prove that a system will operate using the actual input material or detail parts that the real system will have to process, or that its output will be suitable for later operations. These are the questions that a properly conceived prototype will answer, and which must be answered before the production system can be built.

The need for and advantages of prototype systems are not limited to manufacturing equipment. Dearnley and Mayhew[3] found that this approach works equally well in the development of large software systems for business automation.

Another basic principle of man-tech program management is never to make any choice before it must be made. The selection of the supposed production system at the very beginning of phase I was not only a decision which should not have been made, it was a decision which did not have to be made. Especially at the beginning of a program, one should not select a path and eliminate alternatives without good reason. The preliminary trials conducted by the four vendors in phase II are a good example of a simple but rational device for examining all the available alternatives. When two of these appeared to have value, both vendors were contracted to build a more elaborate prototype. If the customer's people are involved in these alternative programs, as they should be, even those which prove unworkable have value because of the insight gained. Perhaps an alternative path that is not suitable for a particular project will prove to be the basis of something ideally suited to some other application.

The examination of alternatives and conducting of experiments can be used as a device for avoiding decisions, especially when large sums may ride on those decisions. If, however, the program plan defines the alternatives to be examined, if the tasks have specific goals which must be met, and if the decision points are defined in advance, there is much less chance of conducting nomadic projects. As the project progresses, the number of technical alternatives must decrease. The tunnel vision shown in the planning of phase I, where a single path was rigidly defined at the start, is no worse than funnel vision which allows each alternative examined to spawn more possibilities rather than selecting or eliminating a possible technical approach.

Another important element in the design of an overall scheme for the management of a given project is to design it to succeed. This is not nearly as easy to do in practice as it appears when stated here. Often, the planning of expensive and important development projects is complicated by all manner of irrational political considerations caused by the need to build the coalitions necessary to gain approval. This ability to get the approval of all of the disparate groups and departments whose support is needed for the project to be funded and ultimately succeed is one of the most important skills of a manufacturing technology manager. The most difficult part of this demanding task is to gain the participation and support of other departments without giving up control of either the project plan or the equipment design. Most groups gathered to support a project will immediately degenerate into a design committee, unless one constantly guards against it. The result of this subtle but destructive process is that programs are designed in order to get permission to start, not to have the best chance of succeeding.

The project manager must realize at the beginning that if it succeeds, no one will criticize him for building one more prototype than was absolutely necessary. But if a million-dollar system is developed and does not work, having saved $3000 on the travel budget will not make him a hero. He must rationally decide, using the insight gained from the preliminary project, what overall scheme will give the best chance of success, then build the project that way. He must not trade off the project's requirements for success in order to be allowed to conduct it.

The development program must, from the very beginning, address the problems of production implementation. In later chapters, this very difficult step will be considered in detail, but as a project progresses and the time when the developed manufacturing process must be turned over to the manufacturing department comes closer, it will become progressively more important. Production start-up will be from 25 to 80 percent of the work of these programs. It must be considered from the very beginning.

Production implementation is not one but many handoffs of functions done during the development, debugging, and testing of the production system by the project staff. The operation and scheduling of the new process is transferred to the manufacturing department. The repair and preventive maintenance of the system is transferred to the various trades within the maintenance department. The quality department takes over responsibility for the input and output of the new process. The overall project management scheme selected must consider these eventual transfers of responsibility.

Responsibility for projects must not be sharply divided between a man-tech development department and a man-tech production support department. The space between the blocks is often a battleground. In this no-man's land, one hears the development group telling how their beautiful, proven system is being destroyed by the stupidity of the grubby-handed goons of the production support group. In rebuttal, the production support group complains that they were handed an ill-designed, poorly built, and untested collection of half-solved problems by a bunch of professors who know nothing about the real world of the production floor. If the groups which know the most about the process are behaving this way, one can imagine what the really tough customers in the manufacturing department are saying.

This antagonism can be avoided if those who will have responsibility for the technical support of the production system are brought into the program when the production system is being designed. This will give them a stake in the system, and it usually leads to a better system because of the insight they have acquired in production support. Figure 4.10 shows this concept schematically.

They must, of course, be innovative people with a bias toward the new. If the staff of the production support department are viewed as antitechnical curmudgeons, there is a tendency to try to keep them away from the new system as long as possible. This generally makes the situation worse when they can no longer be kept out, and they team up with the other forces of reaction to fight implementation. Here, too, one must attempt to build a coalition without giving up control of the project. As there is no single solution for every technical question, there is none for every political problem. But if one knows one will be faced with this problem, one should consider trying to gain their cooperation early.

Usually, the production support groups will be very valuable in the design, building, and testing of the production system. In most cases, the problem will be to free the person assigned to the project from his everyday fire fighting duties so that he can get the large blocks of time necessary to do development work. A large block of free time to a good production support engineer, in many companies, is anything over two hours without a crisis. But the person most in demand for his intelligent problem solving on existing processes is exactly the individual who should be involved in the development of the new; both because of his skills and as a reward.

ALTERNATIVE PROJECT MANAGEMENT TECHNIQUES

The application of the preceding principles of advanced manufacturing technology projects is the subject of the succeeding chapters of this book. But before going to that discussion, the question of what to do with ideas which for some reason cannot be developed, although the preliminary study showed them to have promise, will be discussed. This situation may occur when several promising projects are available simultaneously and a choice must be made because of limited resources.

It could also be that the preliminary study showed the development and exploitation of the idea to be outside of the business area in which the company operates. For example, the most promising potential applications for a development could be in another industry, or the development of the idea could involve equipment research and development that is beyond the ability of the company in which the idea occurred.

Such potential programs should not be dismissed without investigating possible techniques for eventually seeing them in operation. If the project is thought to involve research at a more basic level, before it is ready for development for a specific application, it might be possible to arrange a joint program with a university. Such programs provide extremely valuable interaction between practicing and academic engineers and scientists and students. Besides having potential tax advantages for the company, they can often be conducted very inexpensively, because student labor is cheap and university overhead and administrative costs are much less than most industrial organizations.

The planning of such joint programs often requires special care in safeguarding the company's proprietary rights from the sometimes overly altruistic compulsions and publication mania of university faculty members. These programs must, of course, be planned just as rigorously and rationally as any other. The control of the project to the agreed plan and schedule can also present special problems of diplomacy. But university programs have one thing which has unfortunately almost disappeared from industrial development projects in recent years: They are fun. Although this is something best not admitted to the Calvinists in the finance department, the intellectual stimulation and opportunity to escape from the mental incest that afflicts some companies can, in itself, almost justify them. The opportunity to discover bright young potential employees is another advantage.

Another technique for exploiting an idea which the company cannot, for some reason, itself exploit is to license it to a company that can. Often this is the best way to avoid getting into the manufacturing equipment building business unintentionally. A first-class preliminary study with solid results that provides sound answers to the central technical questions is the only basis upon which such a license can be sold. This is hardly surprising because it is truly the only basis upon which any project should be built.

SUMMARY

- Do not base any project on the unsubstantiated opinion of an expert.
- Believe the results of rigorously planned and conducted preliminary studies and base programs only upon them.
- Do not divide systems or projects unless plans are made and resources are allocated to solving the interface problems that always occur.

- Assign specific minimum technical goals to each of the tasks of a project and use only these to evaluate its success or failure. Evaluating progress by factors such as paperwork processed or man-hours of effort provides no real control because these things do not measure technical progress. In a technical project only technical progress matters.
- Do not expect equipment warrantees to reduce project risk. Only technical insight and sound management reduce project risk.
- Do not base program plans on confidence that nothing will go wrong. Base them on the fact that development work is the sequential solution of problems until (1) none remain, (2) a problem is encountered that cannot be solved, (3) for some reason one stops doing development work.
- Remember that although the final product of a development program is a manufacturing process the equipment for which may be procured from outside sources, development and procurement are entirely different procedures requiring entirely different techniques and must not be confused with each other.
- Plan and conduct programs so that the information needed to make a decision is generated before the decision must be made. If this procedure is followed correctly, the correct decision will, in most cases, be obvious. It must then be made and acted upon.
- Examine alternate technical approaches, but do it so that the program becomes more, not less, focused as it progresses.
- Build prototype systems to solve the central technical questions and test them extensively. Then build the production suitable system. A system designed to be both will probably be neither.
- Prepare honest cost and time estimates based on a plan designed to give the project the greatest chance of success by the most efficient path. Defend those estimates ruthlessly and do not accept a program to which insufficient resources are assigned at the start.
- Do not swap control of a program for permission to conduct it. Use any means necessary to keep important decisions from being made by committees of the uninformed.
- Remember that the goal is an implemented advanced technology manufacturing process. The overall program strategy must specifically consider the obstacles to production implementation and include plans for overcoming them.

Nothing happens unless someone makes it happen. In the planning of a project, *everything* needed for success must be defined and some one person must be given responsibility for assuring that it is done. *Everything*.

REFERENCES

1. Steinberg, B. The military boost to industry. *Fortune Magazine,* 42–46 (April 30, 1984).
2. Anonymous. Miles of laser welds made automatically. *American Machinist,* **127**(11):41–42 (Nov. 1983).
3. Dearnley, P. A., and Mayhew, P. J. In favor of system prototypes and their integration into the systems development cycle. *The Computer Journal,* **26**(1):36–48 (Feb. 1983).

5
THE PLANNING OF MANUFACTURING TECHNOLOGY PROJECTS

INTRODUCTION

A plan is a best-guess model of the future. The accuracy with which it depicts reality, and therefore its usefulness, will depend on the accuracy of the information used to prepare it and the suitability of the model's structure for its intended purpose. For example, there are various models of molecules, each useful for the purpose for which it was designed but unsuited to others. Gas pressure and volume problems are best solved using a model which assumes that molecules are perfectly elastic balls which bounce around infinitely in a closed vessel. Although suited to such problems, this model is useless in the solution of organic chemical synthesis problems; there the Tinker Toy construct is used, showing how the individual atoms are located within the molecule. Problems of nuclear chemistry require more abstract models. Rather than a physical depiction, a mathematical analogy is used.

All of these, whether bouncing balls or Bohr's quantum mechanical model of the hydrogen atom, are just that, models. They are not reality. If one could actually see a molecule, it would not look like any of these things. The reality of the molecule is different from the model used to depict it, so that we may mentally manipulate it and communicate about it. The validity of a model is a function of how well it matches the experimental evidence it is used to explain and predict.

To be useful, a project plan must be connected to the reality it depicts by experimental evidence. The basic information about the central technical questions of the project, acquired from the preliminary experiment and feasibility study, provides this essential link between the model and reality. Without it, the plan is a baseless guess, and PERT charts and critical path analysis and other methods of constructing program models are only devices for hiding technical ignorance behind jargon.

How the plan is constructed and the emphasis placed upon the various elements of the real project by this model of the project will have great effect on the success or failure of the project. If the plan considers only the steps to technical completion but does not provide honest and accurate time and cost estimates, the program may eventually be a technical success but an economic failure. If the planning is only concerned with justifying the most optimistic time and cost estimates, a technically and economically promising program may be prematurely terminated because of apparently excessive time and cost overruns, or it may be a technical failure because sufficient resources were not provided to solve its CTQs. The plan must be constructed to provide the most accurate and most useful model of the project whose reality is still in the future and assuredly will not be identical to the plan. A good program plan is therefore one which provides an honest, balanced, accurate model of a program which will succeed; one which will meet its stated technical, schedule, and economic objectives.

A successful program meets all three goals; on time, on budget, technically successful. But a program may be late or cost more than planned and still be a profit-making asset to a company. If it is not a technical success, no matter if on schedule and on budget, it has failed, and it is a dead loss to the company. The meeting of technical objectives is of overriding importance. This must be kept constantly in mind because the modeling techniques used to develop project plans often overemphasize the importance of financial and schedule factors. However, most technically failed projects are also over budget and far behind schedule. Lack of skill in the planning and conduct of a project invariably leads to total, not partial, failure.

Preparing the project plan permits one to conduct the project on paper as many times as is necessary to define and optimize the most logical path from start to finish. This path is divided into specific steps which are the basis for estimating the time and resources necessary and requesting and allocating those resources. It is also used to estimate the potential competitive advantage expected from the project and to decide how authority and responsibility will be divided among those who will conduct it. Comparison of the planned program with the real program is the basis of project control. It is here that the major differences between the plan and the actual project are usually most apparent. Because plans are too often designed only to provide schedule and cost estimates, the elements upon which these estimates are based are, of necessity, emphasized and technical considerations are treated as secondary. This is particularly true if a planning technique is used which is only suitable to programs where the answers to the CTQs have long been common knowledge. The primary emphasis in the conduct of a development program must be the solution of technical problems. One has only to read the original proposal and the final report of most projects to see this shift in emphasis. This misrepresentation of reality by the program plan can be avoided by the use of a planning technique which is based upon defining the CTQs and the steps necessary to answer them and then implementing those answers.

If the concept of the plan as a model of reality is not understood by all those concerned, the plan can be viewed as a contract and any results which conflict with the plan as a breach of contract. In such situations, the tendency is to ignore or suppress experimental results which might disrupt the plan. The only real path to project success follows the route of technical reality. It is this path which a development project is conducted to discover. Ignoring technical reality in order to follow a planned path, only because it was planned, is to ignore the very information the project is being conducted to develop.

The fastest and least expensive way to do anything is to do it once, right. The plan will develop the estimates of resources which will define the economic desirability to the company of the program. Often the pressure during planning is not to find out how much it will cost to do the program in the most efficient manner, so that the resulting gain to the company is maximized, but to determine how cheaply the absolute minimum successful program might be conducted. If the plan attempts to avoid as many legitimate and necessary costs as possible, the program will be forced to ignore this unrealistic plan by overrunning the schedule and budget. If it does not ignore the plan, it will not perform necessary but unbudgeted tasks. This will lead to technical failure and the waste of all the resources committed to the project.

If a new manufacturing system is developed and installed only to be the last such installation in the industry, the company has not improved its competitive position. The plan must seek the fastest possible implementation in order to gain the greatest lead time advantage. There are those who will argue that it does not pay to be the first to adapt a new technology, but no one has found any compelling reasons for being last. Once it is decided to spend the money needed to develop and implement a new technology, it is indefensible to spend any more time than necessary to reach that goal. The program plan must be designed to get the job done well, quickly, in order to obtain the maximum benefit from the investment.

During project conduct, the plan will be the basis of project control. The plan must therefore provide definite schedule, financial, and technical goals. Financial and schedule goals not linked to technical results are meaningless and only monitor if money and other resources are being spent slowly enough or quickly enough. Cost and time goals must be linked to interim technical goals which show that some useful result has been gained from expenditures.

The plan must have flexibility built into its structure. Sensible flexibility is tenacious, yet it knows that when the impossible, the overly expensive, or the excessively complex is encountered, another approach must be tried. This requires discernment on the part of those conducting the plan but is facilitated if the plan has a built-in mechanism for detecting its own need for modification. This is another reason why specific financial, schedule, and technical interim goals are required to detect when replanning or, in the worst case, cancellation, has been made necessary by experimental reality.

PERT, CPM, AND SIMILAR PROJECT PLANNING TECHNIQUES

It has been shown[1] that the use of performance evaluation and review technique (PERT), while it improves schedule performance in research and development projects, has no effect on technical performance. This is not surprising, because such techniques, whether performed manually or with the aid of a computer, are only schemes for displaying and manipulating information. They do not produce information and certainly do not produce technical insight.

Most program planning techniques are not really ways to develop a plan, but ways to display it. Their suitability for programs in advanced manufacturing technology development and implementation is, at best, limited. This is because the model they produce is usually distorted by overemphasis on clerical and administrative tasks and on the financial and scheduling aspects of the project. These flaws in standard planning techniques result because they attempt to develop a single program model structure which will fit the building of a tract house development and the discovery of a new genetically engineered drug equally well.

Most common program planning schemes are based on a work breakdown structure developed by listing the required tasks and then repeatedly subdividing them until a subtask level is reached at which the subtask is completely understood. For example, suppose a house is to be built to a new design. The first-level tasks in building the house may involve things which have never been done before, for example, building a nine-sided gable. But once this task is subdivided into framing and installing windows and electrical wiring and roof shingling, no mysteries remain. The task may be cut up into completely understood pieces.

In the development of advanced manufacturing technology, as tasks are subdivided, the subtasks become more, not less, novel and mysterious. At the general level, a task may appear quite simple; design and build a machine or an assembly line using a basic concept not applied before in exactly this way. It is only when specifics are reached that the central technical questions and the often contradictory requirements for their solution begin to be apparent. In this case, if standard planning techniques are rigidly applied, it may be necessary to ignore the technical questions the project is supposed to attack in order to produce an acceptable plan.

It is not easy to develop a useful plan while faced with the unanswered questions which, by definition, exist at the beginning of a true development project. If the inescapable uncertainty is not dealt with in the method selected for planning the program, an overly rigid and unrealistic plan will be produced. Then, when the project is being conducted and reality can no longer be ignored, the plan may be abandoned and all control of the project lost.

If PERT or CPM or some such system is a requirement, a plan generated by the method described here can readily be translated into them.

DEVELOPMENT OF A POWER TURBINE SHAFT FOR THE T-55 ENGINE

The T-55 core engine design is used in both a turboshaft series of engines to power helicopters and for marine power plants and in the ALF 502 turbofan engine series to power fixed-wing aircraft (Figures 5.1 and 5.2).[2] This core engine has a seven-stage axial compressor and a single-stage centrifugal compressor, both driven by the same two-stage gas producer turbine system. Aft of the gas producer turbines is a two-stage power turbine system connected to the same power shaft but not connected to the gas producer turbine system. These are front drive, free turbine engines, and power is transmitted either to the fan module or the output gearbox by the shaft to which the power turbines are attached.

The power turbine shaft is about 34 inches long and rotates at over 20,000 revolutions per minute while transmitting the full engine output through the various environments of the engine. These range from the high-temperature gases of the turbine section to the relatively cool, oil-filled, front reduction gearbox (Figure 5.3).

The original engines using this core design were all turboshaft engines and the power shaft and disk assembly was comprised of separate details splined and bolted together. This was necessary because no single material is suitable for the various functions the shaft must perform and the diverse environments in which it must perform them. A material with the high-temperature stress rupture properties required for a turbine disk would not be suitable for a drive spline, even if it were possible to produce an acceptable forging from which to machine the entire shaft.

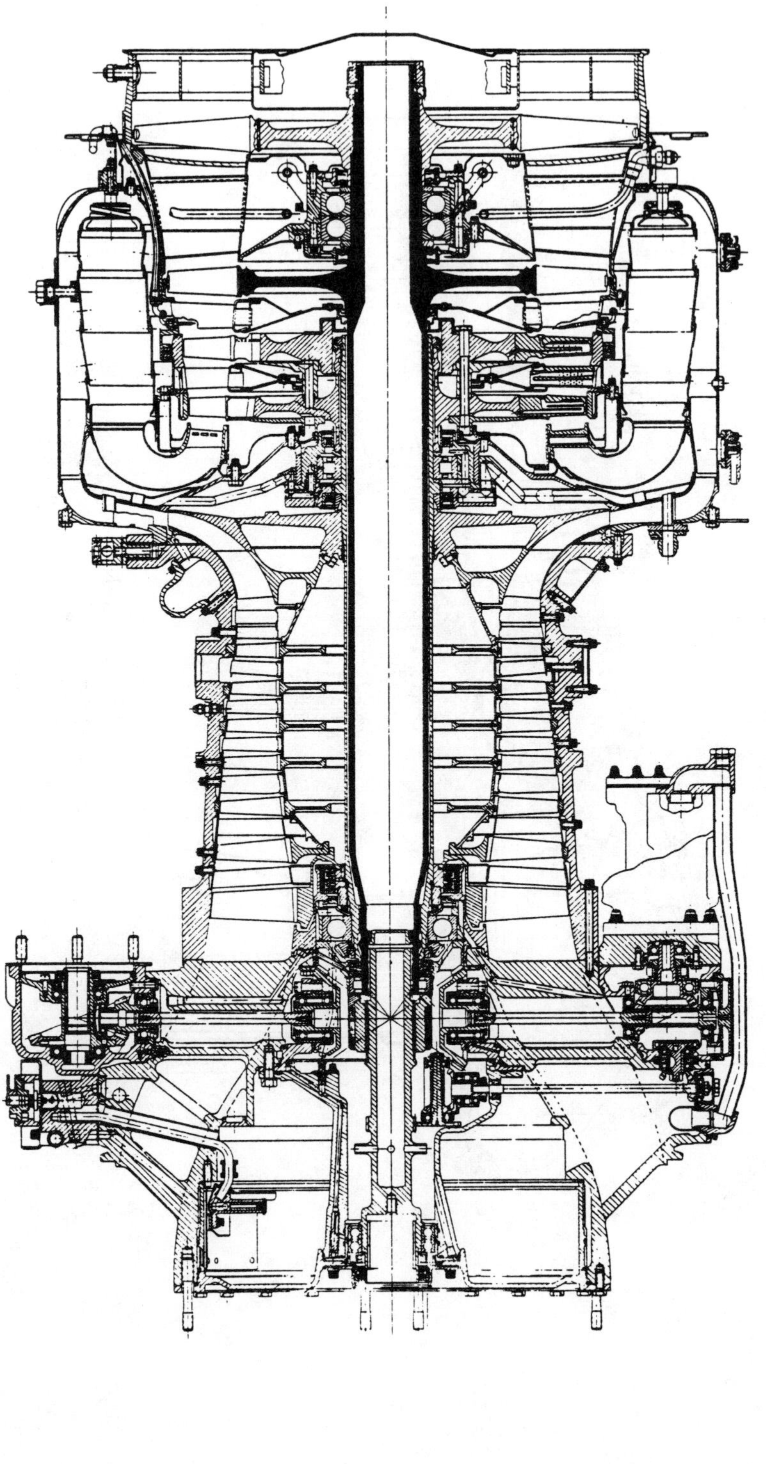

Figure 5.1. T-55 Turboshaft Engine.

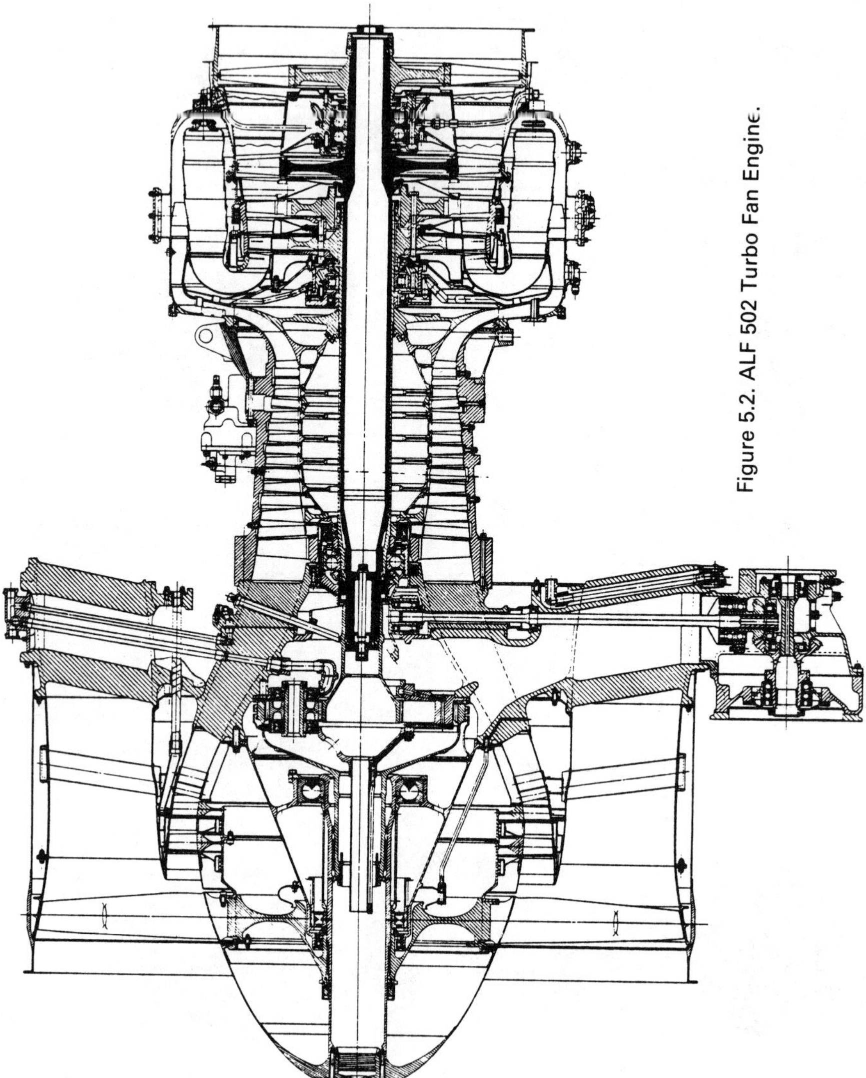

Figure 5.2. ALF 502 Turbo Fan Engine.

Figure 5.3. Power Turbine Shaft.

The metal grains must flow along the lines of stress in order to assure the consistantly high strength required of a turbine disk, yet they must flow lengthwise in a shaft for the same reason. A forging technique which would produce the desired grain pattern in the disk portion of a one-piece forging would give an unacceptable pattern in the shaft portion, and vice versa. Therefore, when the engine was originally designed, the bolted and splined multiple-detailed design was the only way this component could be made. Many turbine engine power shaft/turbine disk assemblies are still designed this way.

Bolts are not nearly as dependable a joining technique as commonly assumed. One never really knows how tight a bolt is, even if the most precise torque measuring devices are used, because its tightness is the result of the load placed on the head minus the friction at the threads and under the head. The exact type and amount of lubricant applied and the exact clearance between the thread surfaces are difficult to control, as is the exact clearance between the hole and the shaft of the bolt. Fairly elaborate techniques such as safety wiring and various locking devices have been developed to keep bolts from loosening in service. But even if the bolt is not actually able to fall out, it can cause the failure of the structure it joins by several mechanisms. A bolt can be overtightened so that it is preloaded or actually stretched during tightening. Then, when service loads are applied, it fails. Or the bolt, although locked so that it cannot loosen further, can be just loose enough to allow the parts to flex very slightly while running, so the bolt shaft moves in its hole and wears and roughens the hole wall, producing areas at which fatigue failures can originate. In the case of a component such as the disk/shaft assembly, which is quite long, operates at very high rpm, and therefore must be very precisely balanced to eliminate engine vibrations, these problems of bolted assemblies can become unsolvable in the real world.

Although each of the machining and assembly problems of a bolted assembly have solutions which will theoretically eliminate them, no way exists to inspect the assembled component to assure that each of these controls was correctly applied. Many factors are involved. The difference between too tight and too loose may be very small, and machining, heat treatment, and inspection of the bolts and the components themselves are complicated. Not only the original manufacturer but various subcontractors and overhaul facilities are involved. Therefore, the chances of an error or an accumulation of several almost-errors are great.

Because of these basic problems in assuring the quality of a bolted joint, service problems occurred with the original shaft assembly. There were, of course, short-term fixes put in place immediately to pile controls, checks, and double checks on every phase of the manufacture and assembly of the components. But it was unpleasantly obvious that, unless some basic changes were made in the disk/shaft assembly's design, no significant long-term improvement in reliability was possible, particularly as more and more powerful models of the engine placed greater service demands on the part.

Richard N. Foster[3] states that when performance is plotted against effort in the development of a technology (Figure 5.4), technical progress is greatest for a unit of effort toward the center of the development cycle and least at both the beginning and the end. The initial generation and development of an idea may, as in the case of the aircraft quality bolt

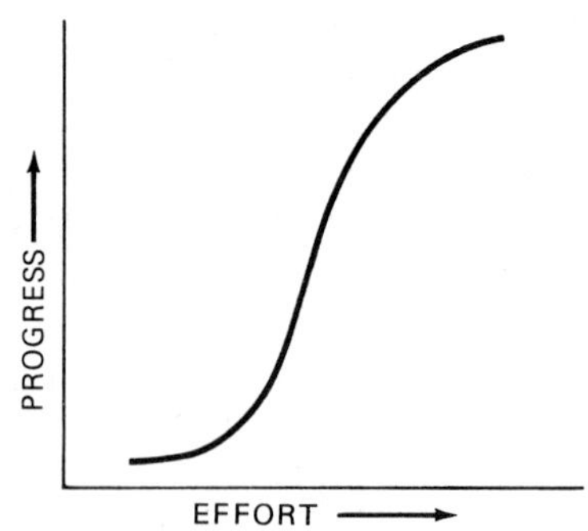

Figure 5.4. Progress vs. Effort Over the Development of a Technology.

from the primitive wedge, take many thousands of years. But there comes a takeoff point above which great progress is made for relatively little effort. This point is inevitably followed by a time when further improvement requires greater effort. Also at this time, technical elegance becomes lost in complexity, as smaller and smaller performance improvements require more and more gadgetry built to tighter and tighter tolerances. The final generation of large reciprocating aircraft engines provides an excellent example of a technology reaching its inherent limit.

The answer to this problem is not to pour more and more resources for less and less gain into a dead-end technology, but to put that effort into bringing a new technology to the point of rapid development. Figure 5.5 shows this concept. When reciprocating aircraft engine development was ended by the shift to jet engines, this is precicely what happened. It is obvious that if all of the effort put into turbine engine development in the last 40 years had gone into piston engine development, the progress would have been miniscule compared to the gains in performance and dependability which actually occurred.

This point, where only the development of a new technical concept for the power turbine shaft assembly would provide a complete solution to the component's problems, was reached in the early 1970s. As part of the original development of the first in a series of aircraft turbofan engines based on the T-55 core design, it was decided to develop an all-welded one-piece assembly made from four separate details of three different materials. This would require the development of the welding techniques for the bimetallic joints that would result, the integration of the radically dif-

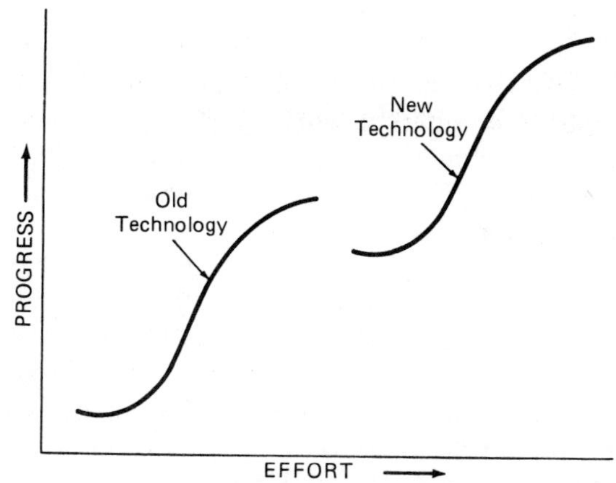

Figure 5.5. The Step to the New Technology Curve.

ferent heat treatment requirements of the various materials, the development of the necessary machining and inspection techniques, and the integration of all of these into an economical, efficient, and inherently controllable manufacturing process. All of the equipment, tools, and fixtures would be developed and either built or bought. The new disk shaft assembly must then pass extensive tests. The entire technology of the shaft's manufacture must then be implemented in a production facility where it would represent a significant technical step for both manufacturing management and labor.

This project was successfully conducted, and shafts to the new design have run many millions of hours without a single service difficulty. It contained all of the elements which increase the difficulty of manufacturing technology development and complicate its management. The product was a critical component of an aircraft engine. Not only advanced technology operations, but an entire process with several such operations integrated with existing technology was required. The new component could not be made without the new manufacturing process, so there was no fallback position.

This project will be used to illustrate the detailed planning of a man-tech program. It is not intended to report exactly how the program was conducted, but how it would be planned if it were to be conducted now. This allows correction of several honest errors that were made in the planning which caused difficulties in the development and the implementation. For example, not enough emphasis was placed on investigating, optimizing, and controlling the standard machining operations which were directly adapted from the process for manufacturing the old shaft. One tends to worry the new high-technology operations to death. Every possible precaution is taken to assure that the electron beam or inertia welding or the ultrasonic inspection is precisely controlled, and everyone double and triple checks to assure that the instructions are understood and are being followed and that the equipment works perfectly. Meanwhile, over in a back corner of the machine shop, a hole of the wrong size is being drilled in every part because no one bothered to check anything so technically trivial as a drill press operation.

It will be assumed that successful feasibility studies have been run to show that it is possible to perform the operations that are central to the success of the project. The shaft design which the various man-tech feasibility studies showed to be best consisted of four detail parts (Figure 5.6). The disk and aft stub shaft were both of a nickel-based superalloy. The central tube was of an alloy steel, and the front stub shaft was of a carburizing steel. Each of these has the properties required for its function and the environment in which it must operate.

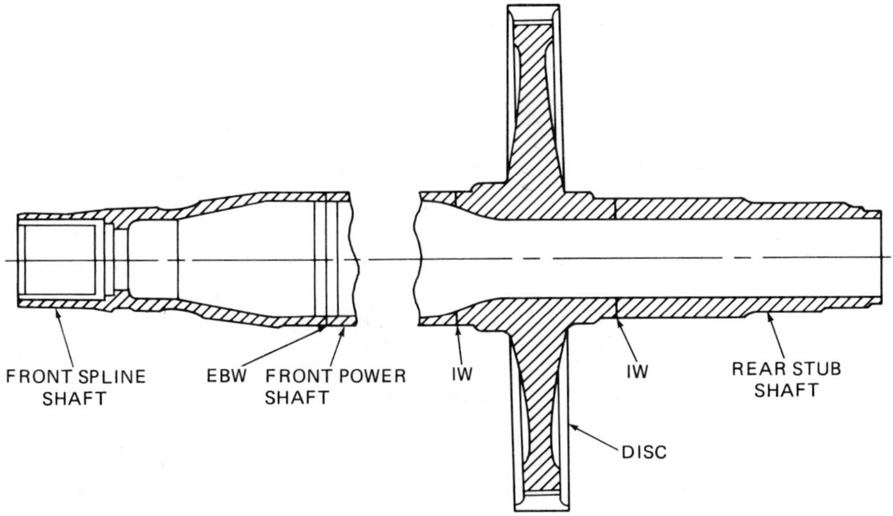

Figure 5.6. Welded Power Turbine Shaft Schematic.

The rear stub shaft will first be joined to the disk by inertia welding. The weld will be machined and inspected and then the disk will be semi-finish machined and the blade tennon slots broached before the assembly is heat treated. This sequence is necessary because machining must be done before heat treatment because of the extreme difficulty of cutting the superalloy after heat treatment.

Next, the front tube will be inertia welded to the disk, then machined and inspected before the assembly is again heat treated to produce the required properties in the tube. This heat treatment is at a lower temperature which does not affect the disk or rear shaft. Finally, the front stub shaft will be electron beam welded to the tube. Electron beam welding was selected for this operation because the necessary heat treatment and machining sequences cause the two subassemblies joined by this final weld to be essentially finish machined. The weld must therefore be made with practically no distortion, maintaining very close tolerances. Electron beam is the only welding process that can weld materials of this thickness while meeting these requirements.

Immediately after EB welding, the weld must be tempered at about 850°F. This operation is complicated by the need to keep an adjacent carburized bearing journal below 350°F so that its hardness and wear resistance will not be reduced. A special heat treating machine will have to be designed and developed and a production suitable version built as part of the program.

QUALITY ASPECTS OF PROJECT PLANNING

A question that must be addressed at the very beginning of planning for an advanced manufacturing development project is the required quality of the product to be made by the process. Without a definition of an acceptable product, it is impossible to construct a plan, because neither the criteria for ultimate project success nor the criteria for the individual tasks can be defined. Also, the individual operations in the process cannot be decided unless the degree of dependability required of them is known. Finally, the quality requirements must be specifically defined so that the development of the required quality control technology can be included in the development project. Too often, projects begin with vague promises that the product will be of matchless quality and end with attempts to find just how bad can it be and still be used or sold.

Product quality is not something that can be ignored until the end of the project and then be obtained by hanging inspection operations onto the process like ornaments onto a Christmas tree. The process developed, and the individual operations which comprise it, must be inherently dependable when operated under the particular conditions and constraints it will meet in the particular shop. The ratio of rejects to total parts produced will not be constant as the production rate of an operation is increased but will, in most cases, increase very rapidly if a certain production rate is exceeded. An important function of a man-tech development project is to find this point for the operations and equipment under consideration and assure that it is not exceeded. The plan must consider this problem and not expect it to be miraculously solved during implementation.

The operator and maintenance dependency of the equipment must be determined for the particular application. If it is necessary for an operator to take extraordinary care and for the equipment to be in perfect condition for the product to be just barely acceptable, either the product or the process or both must be redesigned.[4]

If this redesign is not done during the development project, rationally, by those who understand the problem, it will be done later, when production is beginning, often irrationally, by those who do not understand the problem. Quality is always a central technical question which must be solved, not ignored, by the development project.

The development of the necessary inspection technology must be included in the development project. In the case of the power turbine shaft, this included not only the methods and equipment for ultrasonically inspecting the welded joints, but also the selection of instrumentation to assure that each of the operations was suitably controlled. For example, the temperature monitoring and recording equipment of the specialized heat treatment equipment developed in the project was a basic part of the equipment design. In order to select this equipment, the limits within which the temperatures had to be controlled had to be known.

DEFINITION OF DEVELOPMENT PROJECT OBJECTIVES

All advanced manufacturing technology programs must be planned in detail if they are to be logically justified, conducted, and controlled. However, planning does not assure that a project will succeed and that a competitive advantage will be derived from it. A stupid project, carefully planned and carefully conducted, is still a stupid project.

A feasibility study will, of necessity, be fairly narrowly aimed at finding if the most central technical questions about an idea have answers from which a production suitable manufacturing process or operation can be developed. The exact way in which this information will be used to produce the most advantage for the company must first be decided. This decision will be the basis for stating the overall objective of the full-scale program.

The specific application of the new process must be defined. In order to make something, one must know exactly what is to be made. Even if the new process is to have hundreds of applications, the development program should be focused upon some manageable number of specific parts. These should be selected to represent a reasonable sample of the most difficult tasks the new process is expected to perform. The specific product of the new process must be defined at the beginning of the development project in order to provide tangible goals for the program. "Obtain a robotic welder and use it to reduce manufacturing costs," is not an objective upon which an acceptable project plan can be built. "Develop a robotic welding system (with all required tools, programs, etc.) for welding the X,Y,Z housing more efficiently, and implement it in production," is the correct way to state the program objectives. Even though the housing may be the first of many parts to eventually be made by the new system, it provides the necessary focus without which no measurable objectives can be defined.

Without a specific application, it is not possible to evaluate the economic advantage of the project. Certainly, someone might concoct cost figures to show that a welding robot, on average, may have 85 percent arc-on time, while a human welder will actually be welding only about 40 percent of the time. But without a specific application, such numbers are meaningless. For example, the setup, loading, and unloading time can be estimated only if the actual parts to be processed are known. It may be found that in practice these reduce the actual arc-on time of the robot to 20 percent, because of its inability to cope with part-to-part variations for which a human welder would automatically compensate. It may also be found that the robot's inflexibility to variations in the actual detail parts increases the precision required of other operations and thus their cost. Knowledge of these factors is essential to the economic evaluation of any proposed project, but without specific applications upon which to base them, these estimates are only guesses.

Once the specific application of the technology upon which its implementation will be based is decided, the entire process for making that component must be evaluated in terms of the new operation or operations to assure that the new technology is utilized so as to derive the maximum advantage from it. There is no point to improving an operation which can be eliminated or combined with other operations, if the maximum advantage is taken of the new technology. One sees frequent examples of this turbocharged oxcart technique of improving only one operation or one facet of an operation without reviewing and simplifying the entire process (Figure 5.7). For example, suppose a new machining technology permits higher cutting rates, improved precision, and better surface finishes. If the original part process is based on the sequence of preliminary heat treat, rough machine, finish heat treat, and then finish grind, and the development project is aimed only at directly substituting the new machining technique for the old, the potential payback from the program is greatly reduced. The scope of the program should be enlarged to develop techniques for the new process to machine fully heat-treated material, so that the rough machine operation and the preliminary heat treatment can be eliminated. The new cutting technique thus replaces the original finish grinding operation in the manufacturing process. Combining and elimi-

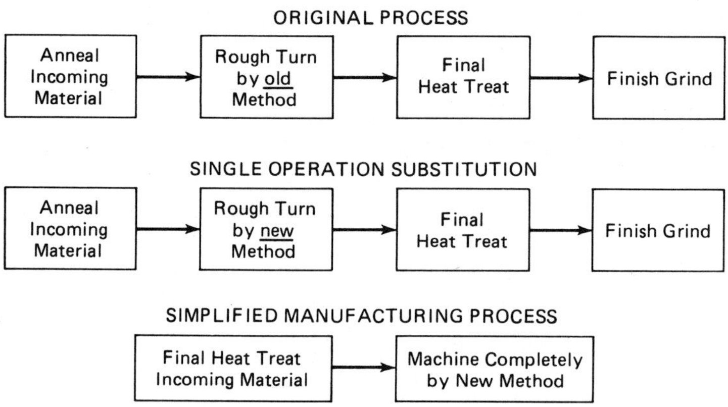

Figure 5.7. Use of Improved Operation to Simplify Entire Manufacturing Process.

nating operations usually produces cost reductions many times those of just increasing the productivity of a given operation. When defining the overall objectives of development projects, this should be constantly considered.

The overall objective of the program should be defined so that the technological step is neither so small as to be trivial nor so great as to be impossible to implement in the particular company. The size of the possible step will depend upon the attitude of the company toward new technology and its familiarity with the particular technology proposed. For example, suppose the program is aimed at replacing punch pressing of various sheet metal blanks with a computer-controlled laser cutter. Further suppose that the company already has in operation a computer-aided design system. One might be tempted to include the development of the necessary software linkage to permit the computer-aided design system to produce the individual part programs for the cutting machine controller. The question is, therefore, whether to first implement the laser cutter using manually written part programs and then, in a later project, develop the necessary post processor and linkages to implement the full CAD/CAM system.

The path chosen will be dependent upon the particular company and the department within the company in which the implementation is to take place. If this were the first advanced system to be installed in a department where the acceptance of the cutting machine itself would be a problem, one obviously should not add the burden of solving all of the manufacturing department/engineering department interface problems involved in a full, one-step project.

If, however, the company has several other full CAD/CAM systems already in operation, one should not needlessly prolong or complicate the project by breaking it into two steps. Deciding on the size of the technological step which a company can make at a given time requires an honest evaluation of the present skill level of those who will develop the technology and those who will eventually operate it in production.

When defining a project's objective based on the results of a feasibility study, one should be suspicious of conclusions generated by extrapolating results and scaling up processes based upon extrapolation. In its extreme form, this can lead to the development of paint rollers 40 feet long. Systems and concepts which are elegant and efficient in one size range often have a point above which they become unwieldy and overly complicated. In doing the preliminary planning for a development project, one should make sure that the objectives are within the capability of the selected technology and that its maximum scale-up point is not exceeded.

If possible, the development project's objectives should be defined so that the advantages of modularity will result from the project. If it is to implement a robotic paint spray system, for example, the future use of components and software of the system for other, similar operations such as burring or polishing should be considered during their design.

P/T SHAFT OVERALL PROJECT OBJECTIVE

As in a feasibility study and preliminary experiment, the overall program objective should be stated in one sentence: "Develop and implement a complete manufacturing process to economically and dependably produce a one-piece power turbine shaft as described by Engineering Drawing XXXX and meeting the minimum performance and quality requirements defined in Specification YYYY."

The drawing and specification will define the specific technical criteria for acceptability of the product and success of the project. At this point, at the completion of the feasibility studies and the start of the planning of the full development program, these documents probably do not yet exist in finished form, but they must obviously be essentially complete before conduct of the program can begin.

The word *economically,* too, is at this point not specifically defined. One of the things which the plan will generate is the estimate of the cost of the project and the cost of producing the shaft. The expected dependability of the resulting process, just what will be the expected rework and scrap rate at each of the critical operations, is also something which must be developed in the plan. It is as important that estimates of process dependability be reasonably prepared as it is that reasonable cost estimates be prepared. Both can then be compared to actual performance as the project progresses. These goals of process dependability must be set and then met. A project which runs over its budget will end, one way or another, in the foreseeable future, but a manufacturing process with poor inherent dependability may be a drain on resources for many years.

WHO SHOULD BE INVOLVED IN THE PROJECT PLANNING

An important consideration in program planning is that those who conducted the preliminary studies must be involved, so that all their experience and insight are used, not just the part that found its way into the report. The other group who must be involved in program planning are those who will be responsible for its execution. Ideally, these two groups, those who did the preliminary work and those who will do the full-scale development, will be the same people. The overall objectives can be set by upper management, but the detailed planning must include these people. The absolutely worst plans for the most doomed of projects are prepared by senior management people who are no longer up to date on the technology but feel that the more expensive and visible the project, the higher the level at which it must be planned. It is critically important that this paper model of the proposed project be as firmly tied to technical reality as possible if it is to succeed. The basis must be insight into the processes and the product. Those things are very seldom acquired or retained for long in carpeted offices.

Generating such plans and then handing them down to those who are to execute them, preferably on stone tablets, is inexcusably bad management. When the plan is completed and approved, the execution of the plan is supposed to begin. In such projects, an additional step is inserted between approval and execution: negotiation of the changes necessary to give the project any chance of succeeding. These are usually conducted in a Byzantine political atmosphere. Middle managers are caught between functional people demanding correction of the most obvious mistakes and the upper-management planners whom they do not know how to confront. By this time, the resources to be assigned to the project are usually unchangeable, so a compromise plan which, in the worst cases, is not even written, is negotiated. In this plan, conflicts between means and objectives are not settled on any basis other than political expediency.

Yet, the necessity to involve those who must conduct the program in its planning should not be used as a reason to convene large committees of the partially involved and poorly informed to plan lowest-common-denominator projects. Only those who have insight into what must be done and how it must be done should have a say in the planning. Burdening these key people by forcing them to explain all of the background information and then diluting their input is a mistake. Only those who have the insight necessary to understand the project and who will have significant involvement in the conduct of the project should be invited to help plan it.

THE PRELIMINARY TASK LIST

Once the project objective has been stated, a list of the actions needed to reach that objective must be made. This list is based on the understanding of the technical problems which must be solved, from the feasibility study. Its preparation is an iterative process and is best performed on a large blackboard. A simple, three-task program should have those tasks as concisely planned as a program with a dozen subprograms and a hundred tasks and subtasks.

Although the preparation of the task list follows the basic logic of the manufacturing process, it is a model neither of the process nor of the development project. The main divisions of the project should be listed first. In the case of the power turbine shaft program, these will be the major tasks. All must be listed before detailed planning of any single task is begun. This is only a list of those things which must be done to reach the project objectives, in the order in which they happen to occur to those making the list.

The plan must be based on experimental evidence and its resulting insight. Because this is a development project, which has only begun, of necessity the experimental evidence needed to plan all of the tasks in detail is not yet available. The information needed to plan the early tasks should have been provided by the feasibility study. But as one looks ahead, down the path the project will follow, the understanding needed to clearly perceive and plan in detail the later tasks is probably not yet available. Therefore, no task should be planned in greater detail than is justified by the available experimental evidence. It is simply a waste of time, and what is worse, these wild guesses can become an obstacle to logical action once the data needed to define logical action have been acquired. There is a tendency to confuse producing page after page of detailed baseless supposition with project planning. They are not the same thing. The only way that everything needed to plan a project completely at the start can be known is if the project has been done before. If such is the case, it is not a development project. Figure 5.8 is the preliminary task list for the power turbine shaft project.

These tasks are not necessarily listed in the order in which they will be done. Several of them will be conducted simultaneously.

Tasks eight and nine are not development work at all; at least at this point in the program they do not appear to be. It is assumed that an existing manufacturing company has incoming material inspection procedures and acceptance criteria in place which would be perfectly adequate for a new component made by a new procedure. This can be a very dangerous assumption. The new process or product may be sensitive to material characteristics which have never before been considered. It is also possible that a characteristic, although supposedly inspected, was not critical to the old technology and the limits of acceptability have been informally widened over the years. Even if all of the procedures and controls are in place, just having someone who understands the requirements of the new technology actively investigate them will alert those responsible and assure that they will know whom to call if they encounter a problem or have a question.

This approach, accepting nothing without having someone responsible

1. Develop rear stub to disc inertia welding technique and verify.
2. Develop front tube to disc inertia welding technique and verify.
3. Develop NDT techniques for inspecting inertia welds and verify.
4. Develop heat treating techniques for inertia welded subassemblies and verify.
5. Develop EB welding techniques and verify.
6. Develop techniques and equipment for heat treating EB welded assembly and verify.
7. Develop NDT techniques for EB weld and verify.
8. Review incoming material inspection procedures to assure that they are adequate and suitable for the PT shaft materials.
9. Review all machining and related inspection procedures to assure suitability to the PT shaft. Develop and verify any new or modified techniques needed.
10. Design and manufacture full-sized fatique test specimens simulating the PT shaft, conduct tests, and evaluate results.
11. Manufacture prototype PT shafts and oversee their engine testing and evaluate results.
12. Integrate results of all subprograms, to produce a draft production manufacturing process for the PT shaft.
13. Manufacture a production prototype lot of shafts in the production shop and optimize the production manufacturing process.
14. Implement the PT shaft manufacturing equipment process and techniques, i.e., turn over the manufacture of the shaft to the production departments.

Figure 5.8. Preliminary Task List of PT Shaft Project.

to the management of the project check it, is the heart of successful planning. The overall objective is successful implementation of a complete production process, and that objective can be thwarted just as effectively by the existing technology, policies, and procedures absorbed into the new process as it can be by the failure of the new technology. A good plan accepts nothing on faith and is built on the questions, what can possibly go wrong, and how can it be kept from going wrong?

Task nine is another example of pessimism being the best planning basis. It reviews all of the operations, not just the new, for suitability. When spending large amounts on projects that can have important effects on the company's future, do not assume or guess or accept someone's assurances that there has never been a problem with that operation in the past. Find out, verify, double check.

Task twelve collects everything from tasks one through eleven and integrates them into a single, logical, complete manufacturing process. This will not happen unless someone does it. The worst implementation problems usually occur, not with the new high-tech operations, but at the interfaces between these operations and the old technology. The problems of conflicting quality requirements, mismatched production rates, and contradictory process instructions are the stress risers at which the program will fail if positive steps are not planned and taken to eliminate them.

Tasks three and seven develop nondestructive inspection techniques, not manufacturing operations. Especially when critical components are involved, these operations are absolutely necessary and cannot be concocted on a crash basis when everything else is ready to run. Their development must be carried on in parallel with the processes they will control. Their requirements must be understood when the unified manufacturing process is written so that the parts in process will arrive at the inspection operations machined to the correct configuration and in the correct heat-treat condition for the inspection operation.

Tasks twelve, thirteen, and fourteen constitute the production implementation. This is often treated as a separate project, to be conducted when and if the development project successfully produces an acceptable product and the process to make it. However, even if implementation is a separate phase, the development project must still be aimed at production implementation and must constantly consider the hard realities it will then face. The areas where this reality is most difficult to confront is in the inherent dependability of the new process, the ease with which it makes a good part, and the difficulty faced in making a bad part. In the worst cases, undependable processes are reported as successful, with the intention of solving the remaining problems during the production start-up phase. Because implementation is much more difficult than development of a new process, this is a mistake. Even if the new process is perfect, more than enough problems will occur in implementation. Development must be complete before production start-up begins, and the program plan must be designed to assure that no problems are left unsolved. Therefore, task thirteen is the manufacture of a pilot lot of parts. Prototype shafts will be made in task eleven using the various processes as soon as their development has progressed sufficiently. This is not the same thing as manufacturing a pilot lot of parts using all of the processes and equipment in the agreed sequence that will be used in production. The total manufacturing process instructions should be drafted (task nine) before this pilot lot is run and then corrected, in detail, based on its results. Someone must follow the parts at every step and assure that all of the final process information is documented.

Once the preliminary task list is drafted, the magnitude of the project begins to emerge. Now, certain management decisions which will affect the detailed planning must be made. The first, of course, is how soon must the program be completed. This will have great impact on the degree of technical risk which will be accepted in order to expedite it. If the intent is to minimize the costs of failure, then each of the subprograms and tasks should be conducted, one at a time, starting with the one about which the least is known or which the feasibility study has shown will require the biggest advance in the state of the art. Once this work has succeeded and is essentially production ready, then the next most difficult task is begun.

In the real world, few projects are conducted in this way, because if a project were of so little importance that the company could tolerate the resulting delay, it probably would not be done at all. Therefore, some parallel web of simultaneous tasks or subprograms will probably be devised. But the plan must still be based on the principle of going right to the most difficult central technical questions and first developing production-suitable answers to them.

In order to construct a detailed plan, in addition to the amount of acceptable risk, one must also know the resources that will be available. The most important resource, by far, is the people that will be made available to the project. Just deciding to assign so many engineers and technicians to the project is useless; the planners must know which engineers and technicians will be assigned, so that the plan can be reconciled with the available skills. For example, if only one person with an understanding of both inertia welding and electron beam welding were available, the project would have to be planned differently than if several people with the necessary skills were assigned. If the personnel and the tasks are not reasonably matched, one person might be expected to do three difficult and important tasks simultaneously. A good program plan does not expect people to do the impossible routinely. Unless the planners know the skills and limitations of the people who are to conduct the program, this can easily happen.

The answers to the questions of how quickly must the program be completed and what resources will be allocated to it must, of course, be consistent. Surprisingly, this is not always the case, particularly if these two decisions are not made by the same people or are not made by the same people within the same logical constraints. This is why it is best for the person in charge of planning the project to prepare the preliminary task list and then, with this as a basis, try to get these two important decisions made and approved together.

ASSIGNMENT OF TASK RESPONSIBILITY

Once the professional skills available and the target for completion of the program are decided, the task list should be redrafted so that the tasks fit the resources available, and so that a program management structure suitable to the particular project can be developed.

In the P T shaft program, the inertia welding processes were developed by a group reporting to a single engineer, and the electron beam welding was developed by a second group under a second engineer. Responsibility for the development of the nondestructive inspection techniques (NDT) for each welded joint was the responsibility of the engineer developing the joining process, but the work was done by an NDT development lab, performing as a functional group in a matrix structure.

These two engineers or principal investigators both reported to the program manager. Figure 5.9 shows the division of responsibility for the subprojects. (Task numbers are constant throughout all figures.)

Conflict may occur in the NDT lab. If both principal investigators arrive at 3 P.M. on the last Friday of the project, of course, there will be conflict. But this should not happen and, in fact, did not happen. The development cycles of the welding processes were such that the three NDT programs were conducted sequentially, with no two reaching the lab together.

This division of responsibility gives each principal investigator about the same total work load. It assigns the tasks that could go to either, such as manufacture of prototype hardware and its testing and the drafting of the final production process, to a single investigator, although both will have input into these tasks. The plan assigns every task to one or the other of the principal investigators to assure that each knows exactly what he is responsible for completing. If this is not done, these critically important tasks, requiring input and action by both groups, will be only partially or poorly completed. If responsibility for the completion of a task is assigned to two people, it is assigned to no one.

Although responsibility for the completion of each task is assigned to one specific person, this does not mean that only that person does the work necessary or that only his insight and skills are used to perform the task. Assigning specific responsibility does not place tasks or subprojects inside of separate little organizational fortresses into which no outside skills or knowledge can penetrate and out of which no questions can escape. Assignment of task responsibility must not be misunderstood to mean that the project is fragmented into small factions operating independently or, in the worst cases, in competition or at cross purposes with each other. The person responsible for each task is responsible that every

Principal Investigator A

1. Develop rear stub to disc inertia welding technique and verify.
2. Develop front tube to disc inertia welding technique and verify.
3. Develop NDT techniques for inspecting inertia welds and verify.
4. Develop heat treating techniques for inertia welded subassemblies and verify.

10. Design and manufacture full size fatigue test specimens simulating the PT shaft, conduct tests, and evaluate results.
11. Manufacture prototype PT shafts, oversee their engine testing, and evaluate results.

Principal Investigator B

5. Develop EB welding techniques and verify.
6. Develop techniques and equipment for heat treating EB welded assembly and verify.
7. Develop NDT techniques for EB welds and verify.
8. Review incoming material inspection procedures to assure that they are adequate and suitable for the PT shaft materials.
9. Review all machining and related inspection procedures to assure suitability to the PT shaft. Develop and verify any new or modified techniques needed.

12. Integrate results of all subprograms, above, to produce a draft production manufacturing process for the PT shaft.
13. Manufacture a production prototype lot of shafts in the production shop and optimize the production manufacturing process.
14. Implement the PT shaft manufacturing equipment, process, and techniques, i.e., turnover the manufacture of the shaft to the production departments.

Figure 5.9. PT Shaft Project Assignment of Task Responsibility.

source of the skills and insight needed to best perform that task is acquired and used. There is no contradiction in this, and those who will plan and conduct the project must understand it from the start.

Confusion about who is responsible for what can easily happen if the assignment of task responsibility and the project flowchart are not kept separate, and a murky model of how the project will be conducted is allowed to result. Figure 5.10 is an example of this. The use of blocks and arrows gives a misleading appearance of conciseness. Yet, the meaning of the blocks and arrows is not even consistent throughout the diagram. Is each block a task or a function? At the top the arrows imply responsibility, yet toward the bottom they seem to imply the sequence in which the subprograms will be conducted. The two NDT development tasks and

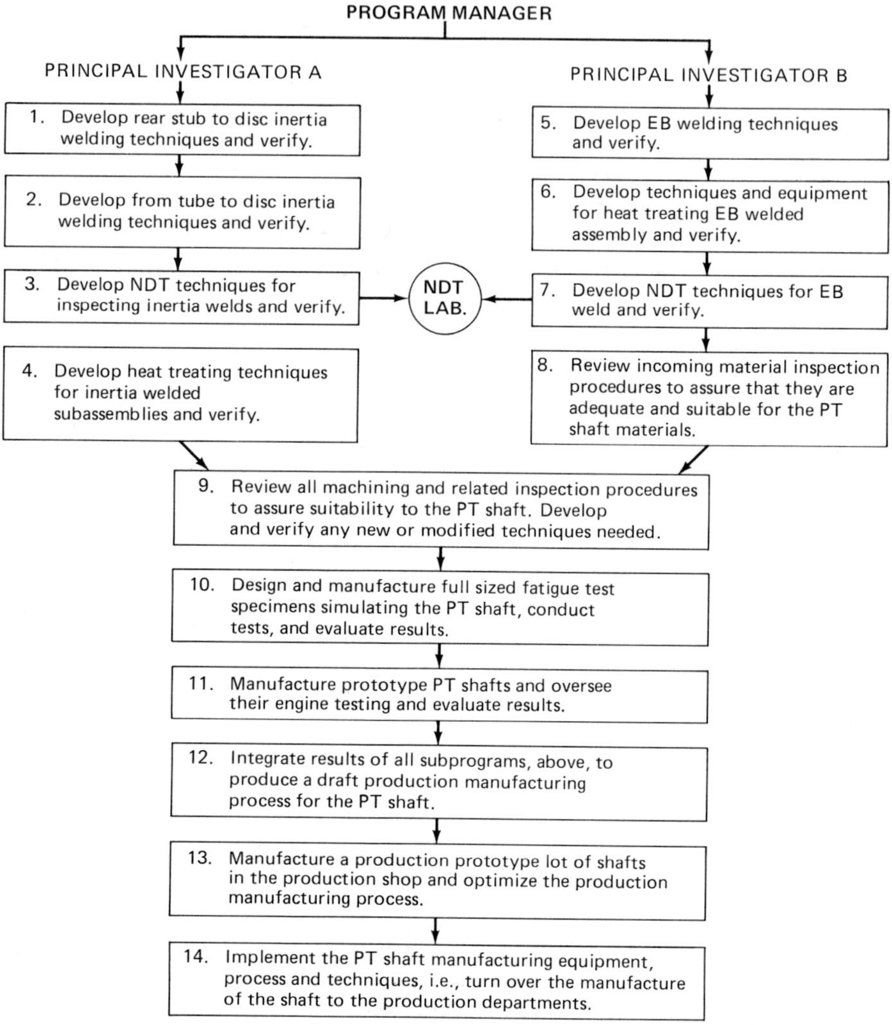

Figure 5.10. Murky Project Diagram Neither Accurately Modeling Project Nor Assigning Responsibility.

the NDT lab are arrowed together, but the exact relationship implied is unclear. Equally unclear is just who is responsible for the development of the NDT techniques. The implication is that it is shared between the principal investigator and the NDT lab.

This, of course, means that no one is responsible, and it is one of the most common causes of failure in matrix organizations. The NDT lab should not even appear on the assignment of task responsibility because they have no responsibility for task completion. Although they will do the work, the responsibility for its completion is the principal investigator's only. If they are not doing the job, he must do whatever is necessary to get them to do it.

The arrows joining tasks four and eight to task nine imply neither responsibility nor sequence, but the flow of information from earlier tasks to later ones.

The task responsibility assignment in Figure 5.9 will only work if the two principal investigators cooperate and if the program manager realizes that his main job is to assure constant communication and to settle conflicts quickly, fairly, and logically. If this scheme of task responsibility is adopted because the program manager wishes to be the only one involved who knows what is happening and will use this information to play one principal investigator against the other, it will not work. But no plan will work well if subjected to such a management style.

Once all of the tasks are listed and responsibility for them is assigned, they should be arranged in the order in which they will be done. Tasks to be done simultaneously, or at least begun at about the same time, should be placed side by side. Figure 5.11 shows principal investigator A's task list so arranged.

Notice that the front tube to disk inertia welding technique will be developed first. This weld is between dissimilar materials with very different high-temperature properties which could complicate their welding, and the location of this weld requires it to operate in a more stringent service environment. The most difficult task must always be done first.

Task three has been divided in two because the development and verification of the welding and NDT techniques are interrelated and must be conducted simultaneously. Ideally, preliminary welding and NDT techniques will be developed together. Then, as samples for destructive weld evaluation are made, these will be inspected by the NDT technique so that the metallography and other testing will verify both using the same specimens. If the NDT technique is available early enough so that less

2. Develop front tube to disc inertia weld and verify.	3a. Develop NDT techniques for front to disc inertia weld and verify.
1. Develop rear tube to disc inertia weld and verify.	3b. Develop NDT techniques for rear tube to disc and verify.
4. Develop heat treating techniques for inertia welded subassemblies and verify.	

10. Design and manufacture full sized fatigue test specimens simulating the PT shaft, conduct tests, and evaluate results.

11. Manufacture prototype PT shafts and oversee their engine testing and evaluate results.

Figure 5.11. Principal Investigator A's Task List in Order of Conduct.

than perfect welds are still being made, the experience gained in finding real defects rather than searching for drilled holes will be very valuable. This simultaneous development technique is superior to a sequential approach because each development task gains important insights from the work being done in the other. Task four must also be conducted simultaneously with tasks 1 and 2, so that the heat-treating techniques are available when needed.

The sequence of tasks ten and eleven is obvious.

Figure 5.12 shows how the second principal investigator will be spending his time. First, the welding technique and the method for evaluating the welds will be developed together because if both cannot be made to work, none of the following tasks are necessary.

Next, the equipment for heat treating the shaft must be developed and built. This job can wait because it involves no unanswered CTQs, and the test welds will be heat treated in a furnace because they do not have the carburized areas adjacent to the welds which must be kept cool.

Tasks eight and nine will be done next because they are dependent on information about the process for the new shaft that will emerge from the work on tasks one through six. Until some knowledge of the new process and its requirements is developed, any attempt to evaluate existing processes and procedures and locate potential integration problems will be useless.

Figure 5.13 shows these two simultaneous subprograms, but it does not really show how they will be time phased so that the information necessary

5. Develop EB Welding techniques and verify.

7. Develop NDT techniques for EB weld and verify.

6. Develop techniques and equipment for heat treating EB welded assembly and verify.

8. Review incoming material inspection procedures to assure that they are adequate and suitable for the PT shaft materials.

9. Review all machining and related inspection procedures to assure suitability to the PT shaft. Develop and verify any new or modified techniques needed.

12. Integrate results of all subprograms to produce a draft production manufacturing process for the PT shaft.

13. Manufacture a production prototype lot of shafts in the production shop and optimize the production manufacturing process.

Figure 5.12. Principal Investigator B's Task List in Order of Conduct.

POWER TURBINE SHAFT PROGRAM TASK LIST

4. Develop heat treating techniques for inertia welded subassemblies and verify.

Subprogram A

2. Develop front tube to disc inertia welding technique and verify.

3a. Develop NDT techniques for front tube to disc weld and verify.

1. Develop rear stub to disc inertia welding technique and verify.

3b. Develop NDT techniques for rear tube to disc weld and verify.

10. Design and manufacture full-sized fatigue test specimens simulating the PT shaft, conduct tests, and evaluate results.

11. Manufacture prototype PT shafts and oversee their engine testing and evaluate results.

Subprogram B

5. Develop EB welding techniques and verify.

7. Develop NDT techniques for EB weld and verify.

6. Develop techniques and equipment for heat treating EB welded assembly and verify.

8. Review incoming material inspection procedures to assure that they are adequate and suitable for the PT shaft materials.

9. Review all machining and related inspection procedures to assure suitability to the PT shaft. Develop and verify any new or modified techniques needed.

12. Integrate results of all subprograms, above, to produce a draft production manufacturing process for the PT shaft.

13. Manufacture a production prototype lot of shafts in the production shop and optimize the production manufacturing process.

14. Implement the PT shaft manufacturing equipment process and techniques, i.e. turn over the manufacture of the shaft to the production departments.

Figure 5.13. Complete Task List in Order of Conduct.

for a given task will be available when necessary. The problems of coordination and communication which will obviously occur are the responsibility of the program manager. He must stay current in both subprojects without unduly intruding on the principal investigator's authority. He can shift resources or put pressure on functional departments as needed, so that project momentum is not lost through schedule conflicts. The worst delay is one which stops both subprojects. The manager must assure that this does not happen.

This task list is a general model of how the project will be conducted if nothing goes wrong. Something usually does go wrong, and it is the project manager's job to recognize and solve problems, hopefully before they occur but certainly before they become serious. One of the primary control functions of the plan during the conduct of the project is to help the manager recognize problems before they become serious.

This is a fairly rational overall plan of the project, in terms of what must be done, in approximately what order. Tasks one through seven must solve all of the central technical questions of the program before tasks eight through fourteen can be done. This does not mean that tasks eight through fourteen are less important to the final outcome of the project or that they can be done less professionally. Sometimes, if difficulties are encountered in the earlier tasks and they cost more or take longer than planned, there is a tendency to recoup these losses by rushing the later, less exotic tasks. This can lead to prolonged and difficult implementation (task fourteen) when the deferred problems these tasks should have solved must be solved.

THE DETAILED TECHNICAL PROJECT PLAN

Once the final task list is completed, the detailed project plan can be prepared for the major tasks. In the turbine shaft program, the first tasks are the front shaft to tube inertia weld development and the development of the NDT techniques for this joint. Although for purposes of assigning responsibility the task list treats these as separate activities, in the actual project they will be closely interrelated. The task list is not a model of the project and should not be mistaken for a model or substituted for it. It is a list of what must be done in what order, not a plan of *how* it should be done. A plan must include specific technical criteria for acceptable completion of each of its steps.

The actual preparation of the detailed plan for performing the tasks begins by listing, vertically, the central technical questions which must be answered by the project, subproject, or task. For the shaft to tube weld development these would be

CTQ 1. Is it possible to produce an inertia weld joining the disk material to the tube material with properties suitable for the P T shaft, and are NDT techniques possible to assure the quality of these welds?

CTQ 2. How can inertia welds suitable for this application be made and their quality assured?

CTQ 3. How can the technology developed in answering questions 1 and 2 be implemented in production, and what problems must be addressed in its integration into the complete manufacturing process?

The first question should have been answered by the feasibility study before the planning of the development program was begun. It is ridiculous to plan an expensive program to develop something without first proving, experimentally, that it is at least possible. The question, although already answered, should be listed along with those to be answered by the development plan. This is to assure that the feasibility study and the development program are logically consistent, and that what is to be developed is actually what was proven feasible.

The question about the welding process and the NDT technology to control it was combined because they must be attacked together in the conduct of the program.

Next, the actual subtasks necessary to answer the CTQs are listed next to them, as shown for the heat treatment development task in Figure 5.14. Next to the subtasks are listed the minimum criteria for successful completion of that subtask. As in the planning of a feasibility study, this allows for flexibility in the methods used to reach rigidly defined goals. This is essential to conducting a controlled yet flexible project.

These three columns, the CTQs, the task descriptions, and the task criteria for success are the heart of the development project plan. Notice that they are only technical, although a substantial part of any program is involved with administrative functions, such as equipment and material procurement, time and cost recording, and so on. The inclusion of these functions in the technical plan often leads to their masking or even substituting for technical progress. For this reason, these things are treated separately in a logistic support plan to be explained below. It must be remembered throughout the planning and conduct of technical programs that technical success is the absolutely essential concern.

These three columns of the plan define what must be done, how it is to be done, and the minimum definition of its successful completion. It is in their preparation, while working out the precise wording of these statements, that the program manager, principal investigators, and any

PT SHAFT PROJECT—I-WELDED SUBASSEMBLY—HEAT TREAT TASK

CTQ	Subtask	Subtask Acceptance Criteria
1. How should the front tube detail be heat treated before welding, and how should the welded assembly be heat treated to give the required properties to the assembly?	1.0 Prepare x number of preliminary tube specimens with 1725°F normalize and 1200°F temper and inertia weld to disc material.	1.0 Minimum impact strength shall be . . . Minimum torsional fatigue strength shall be . . .
	1.1 Temper x/3 specimens at 1200°F, x/3 at 1150°F, and x/3 at 1100°F.	
	1.2 From each lot from 1.1 machine y specimens into impact test bars and test and z specimen into torsional fatigue specimens and test.	

Figure 5.14. Task Plan.

others involved in the program, such as the NDT lab supervisor, precisely define what the program is and how it will be conducted. This planning should be done starting from a blank blackboard, with everyone present, and with no interruptions. If it is done correctly, it will take a great deal longer than expected, and it will uncover all sorts of communications problems and conflicting opinions about precisely what the project is intended to do and reconcile them.

This is, of course, the setting of goals and objectives which is crucial to the planning and conduct of any project. But the use of this method of defining them assures that the overall objectives, the means to be used to reach them, and the interim goals which will define progress toward them are concise, consistent, and understood.

The statement of the central technical questions is the way all of those involved in a task agree on exactly what it is they must ascertain. The question should define this precisely and not be so general and all-inclusive as to be useless. For example, the CTQ of the heat treatment task (task four in subprogram A) could be stated: "What heat treatment of the detail components, subassemblies, and completed assembly are required to impart the best mechanical properties to the P T shaft assembly?"

This question does not clearly define the *central* technical questions, because much of the information it requires to be developed is already known. The heat treatments which impart the best properties to the materials were either available from the start or developed in some earlier program which selected the materials. The disk and rear tube are of the same material and thus present no integration problems; once welded, they will be given the best heat treatment for that material. The problem which must be addressed in this task is to find the heat treating procedure for the welded three-piece assembly of the disk and both tubes which gives the best properties to the front tube and the weld joining it to the disk. The question to be answered in this task is, therefore: "How should the front tube detail be heat treated before welding and how should the welded assembly be heat treated to give the required properties to the assembly?"

Notice that it asks for the required properties and not the best properties. It is hoped that the program will develop technically excellent solutions to its problems, but the acceptability of those solutions is judged by meeting firm minimum requirements, not perfection.

The preferred heat treatment of the tube before welding was normalizing, followed by tempering at 1200°F. Therefore, this was the preweld treatment tried first. The required postweld heat treatment was unknown, but it could not exceed 1200°F without reducing the properties of the tube, so three temperatures were tried. If none of these procedures had produced the minimum properties, a new subtask, based on the insight gained, would be defined as a matter of routine. That is how development work is conducted. However, the CTQs and the minimum acceptance criteria are not subject to redefinition without complete agreement of those who first approved the project. In this way, both flexibility and control of the project are obtained.

This is precisely what happened in the development of the welding techniques for the rear shaft to disk joint. Because both materials were the same and the feasibility study on subsized specimens showed no difficulties, this was assumed to be the easier of the two inertia welding tasks. It wasn't. When full-sized specimens were welded and subjected to ultrasonic inspection, defects were detected. These were eventually found to be caused by the combination of flywheel speed and thrust used in the available welding machine being inappropriate for a joint of this size in this material. It was necessary to repeat the relevant subtasks using an inertia welder of much higher capacity to produce welds which met the strength and quality requirements.

Another example of this method of preparing the detailed technical plan is given in Figure 5.15. This program is the selection and implementation of a computer-controlled laser cutter for sheet metal fabrication. It is to replace punch press, milling machine, and electrical discharge machining on a variety of parts already being manufactured by these older techniques.

PROJECT PLAN—PRODUCTION LASER CUTTER

OBJECTIVE: TO IMPLEMENT IN PRODUCTION A LASER CUTTING SYSTEM FOR THE PARTS LISTED IN THE SUPPLEMENT

CTQ	Tasks	Task Accepting Criteria	Time Schedule	Engr. Time	P.O. Cost	Mat'l.	Travel
1. Is it possible to laser cut slots such that quality cuts & quality braze joints result?	Feasibility Study done.	X	X	X			
2. What laser cutting techniques, machine layout and computer are suitable for application?	2.1 Survey potential vendors and select candidates for trials.	2.1 Must be capable of demonstrating techniques and building a production system.	1 Mo.	80 hr			
	2.2 Set up and cut (3) diffuser plates at several (perhaps 4) candidate vendors.	2.2 Must meet B/P requirements and spec. xxx. Two best to be brazed and tensile tested. Must meet xxx minimum properties. **Selection of Best** Two will be selected based on quality (above) and demonstrated technology.	2 Mos.	120 hr		20 Diffuser plates	

	2.3 Set up and cut housing front plates at 2 best from 2.2.	2.3 Must meet B/P Requirements. spec xxx. (Metalographic included) Best will be selected on quality (above) and demonstrated technology.	2 Mod.	120 hr	10 Housing front plates
	2.4 Evaluate results, select vendor, write spec. for system.		1 Mo.	80 hr	
		Totals	6 Mos.	400 hr	
3. How can system best be implemented in Production and what problems must be addressed?	3.1 Design, build and qualify system.	Successfully cut at least ten (10) jobs (including 2 above) on production basis meeting B/P and spec. xxx.			
	3.2 Implement system including training of *all* personnel.	System routinely meet quality and schedule requirements while run, programmed and maintained by Operations Dept.			

Figure 5.15

Perhaps 50 applications for the machine were found in a preliminary survey, and the 10 parts which were the most difficult or expensive to make by the older techniques were determined. The overall objective of the program was then stated as follows: "Develop an automated laser cutting system for the following listed parts and implement it in production. (Parts list follows.)"

Of the 10 parts listed, the two which were the most difficult to manufacture by the old methods and which appeared to be the most challenging to cut with a laser were selected. These parts were both to be high-temperature nickel brazed at a later operation. Airfoil-shaped vanes would be inserted in the laser cut slots and brazed. The feasibility study therefore had to not only answer questions about the ability of the laser to cut the required materials to the required tolerances but determine that the resulting cut surfaces would be suitable for brazing.

Whether this is a development project may be questionable, since all of the technology needed was being offered for sale by a number of possible suppliers. It could also be called a selection and implementation project. Whatever it is called, such a program is the correct way to accomplish such an objective.

Notice that the project is focused on selecting the technically most suitable system and then solving as many potential implementation problems as possible before the system arrives on the shop floor. The cost of such projects is trivial compared to the cost of the systems they implement. They pay for themselves many times over in the reduced start-up and operating costs of a better system.

THE LOGISTIC SUPPORT LIST

In any technical project, there are many nontechnical clerical and administrative tasks that must be done in order to accomplish the technical goals. These things include ordering, receiving and shipping materials, issuing purchase orders, and searching for all of these things when they are lost in the system. It is important to keep these functions separate from the technical effort in the planning of a development project for several reasons. First, if obtaining the material is included in the plan as a task equal to finding a way to process that material, the importance of the clerical task to the technical success of the project is exaggerated. No one has ever run a development project to find improved ways to get sheet metal out of a supply crib, although in some companies it might be worthwhile. A sure sign of this exaggeration is when "obtain sheet stock from supply crib" appears on the plan as an interim goal.

The second reason for segregating logistic support tasks on a separate list is to allow them to be done by someone other than the technical people whose primary concern must be technical success. In some organizations, the time necessary to fill out requisitions and get the required signatures, make travel arrangements, and justify minor expenses can consume as much or more time than the actual work of the project. If such tasks are not kept separate, technical people will waste time doing tasks that should be assigned to clerical and administrative people. But, more important, the frustration that these trivialities can generate is a strong deterrent to innovative effort. Technical staff should not do clerical work. They probably do not do it very well, anyway.

The logistic support list of the program plan comprises all of these nontechnical but essential tasks. In the adjacent column are listed the technical tasks which are dependent upon the completion of each support task. If technical subtask 2.3 needs specimen material procured, cut, and delivered to a subcontractor, and the same material is needed at another location in subtask 4.3, procure, cut, and ship material is listed once with both subtasks following it. This approach has the added advantage of consolidating these tasks, so that the various subprograms or tasks do not all waste time doing the same logistic tasks separately. For this reason there should only be one logistic support list for the entire project; it should not be divided into subprojects.

Figure 5.16 is the list for the laser cutter implementation project. It contains only those tasks that are strictly logistic. For example, it does

LOGISTIC TASK	DEPENDENT TECHNICAL TASKS
1. Obtain vane diffuser blanks and deliver to all subcontractors.	2.2, 3.1
2. Obtain air diffuser plate blanks and deliver to subcontractors.	2.3, 3.1
3. Issue purchase orders to vendors for preliminary trials.	2.2
4. Issue purchase orders to vendors for final selection trials.	2.3
5. Obtain return of cut and unused materials from all vendors.	2.2, 2.4
6. Obtain required approvals of specification and purchase order for selected system.	3.1
7. Obtain materials needed for development of eight additional jobs and deliver to subcontractors.	3.1

Figure 5.16 Logistic Support List.

not list making telephone calls to prospective suppliers in order to screen them, because this requires technical understanding of the project.

As in the technical plan, those things to be done first are the most specifically described, and the latter tasks are drafted with the understanding that detail will be filled in before they must be done, based on the insight gained from performing the earlier tasks. For example, when the final system is being installed, undoubtedly a number of logistic support items will be added to the list, but until the system is defined, it is a waste of time to attempt to predict exactly what these items might be.

Another advantage of separating logistic from technical tasks, but making both traceable to the solution of the central technical problems of the project, is that it helps to eliminate unnecessary administrative tasks or at least forces them onto some other budget. If a company has any administrative reports that must be prepared periodically for every project, these will turn up on the logistic support list with no technical task dependence. If nothing else, this will show that such reports are irrelevant to meeting the technical objectives of the project and therefore should not really be charged to it.

Provision must be made for providing timely schedule and cost data to those running the project. If the company's normal accounting system actually provides this, there is no problem and the project plan can treat this as a given. In many companies, this is not necessarily so. If one waits for complete, accurate information through normal channels, it will be obsolete when it arrives. If this is the case, the project manager should set up his own system of tracking costs and schedules so that he can detect problems *before* they happen. It is during the planning that an honest assessment of the ability of the company's normal system to provide timely, accurate data, to what are always nonstandard activities, must be made. If necessary, some alternate method must be devised to compensate for system weaknesses.

SCHEDULE AND COST PLANNING

Once the technical plan is complete, the next step is to estimate how long it will take to complete each of the tasks and subtasks. This is not how many man-months it will take, but an estimate of how long the resources assigned will need to complete the task if it goes as planned. These estimates for each task are then summed to find the total program duration. Simultaneous tasks are not summed, of course. In many cases, the total time to complete the program is a given quantity, generated by factors outside the project.

Because of the emphasis placed on the technical success of projects here, one should not assume that meeting schedule and cost goals is considered unimportant. Technical success is the primary consideration, but this does not mean that cost and schedule goals are insignificant. Obviously, a program that would give a company a major competitive advantage if completed for $200,000 in one year might not be a success if the same goals were met in four years and cost $2 million.

If a maximum time to completion is given, the task times must be assigned so that they total this number. Later, when a working schedule is prepared, slightly less time may be apportioned to each task as part of the project control mechanism. But, in the preparation of the original program estimates, the most likely duration of each task and subtask should be used. This is so that a contingency factor will be rationally estimated and added only once. If this is not done, each level of management may add or subtract an unstated factor from the estimates, depending on the political atmosphere in which the project is being planned. This insidious estimate paring or padding is probably the most common cause of unrealistic project plans. An unrealistic plan is no plan at all.

It is here, while making the estimates of duration, that conflicts between the resources assigned and how long it will take to meet the project objectives become obvious. Some tasks take a given time no matter how many or how brilliant the people assigned to them, for example, a 500-hundred hour test. The worst form of this conflict is the project that must be done in a set period of time which the estimates show cannot be done, no matter how the essential tasks are telescoped. The temptation is to eliminate supposedly unessential tasks and hope for the best. If the technical planning has been done correctly, there are no nonessential tasks in the plan, since only those required to answer the central technical questions and implement those answers are included.

A decision to go ahead with a project on a crash basis, intentionally ignoring one or more central technical questions, may possibly be justified in an absolutely critical situation. But I cannot conceive of it. The risk of project failures that such a wild approach generates forces one to ask that if the project were of such overriding urgency, should not risk of a calamitous failure be avoided at all costs, even to extending the schedule?

Often these chaos-causing emergencies are found to be generated, not by the desperate expediencies used to justify them, but by some personal political problem which has festered into a project-upsetting crisis. One of the surest signs that this is the case is when every attempt to negotiate a rational deadline is met with the answer that *they* want it done this way, so it cannot even be discussed.

The estimates must include the time to perform not only the technical tasks but the logistic support tasks. Another advantage of separating these tasks is that the often disproportionate time and resources required by nontechnical tasks thus become obvious. If estimates of the time to perform the project must be reduced, the least harmful cuts can and should be made in logistic support time. If three months must come out of the project, do not take it from the time needed to invent and test something, take it from the time needed to process paperwork. If one is negotiating the total project duration with management, offer to cut the estimate in return for being assured priority treatment in the processing of requisitions and purchase orders, for example.

The time to complete the logistic support items upon which the technical tasks depend should be included in the technical task time estimate if using separate estimates of the times necessary to perform purely administrative tasks makes the plan unduly complicated and confusing. In this case it is best to estimate the time to technical completion of a task, taking into account the time necessary for the logistic support items, and then use whatever force is necessary to assure that no administrative item is allowed to hold up technical progress.

Once the best estimate is made of how long it will take to complete each of the tasks necessary to solving the CTQs, an honest evaluation of just how sound these estimates are should be made. Obviously, if a very similar task has been done before, in the feasibility study perhaps, the estimated time should contain little or no uncertainty. As the nature of the particular task becomes less familiar, the estimate becomes less certain. A complete project plan, based upon a well planned and conducted feasibility study, will contain many more tasks about which something is known than one begun without this necessary first step. Generally, however, the estimates of later tasks will be less solid than those to be performed earlier in the program. The degree of uncertainty in the estimate of each task should be itself estimated and this estimate used as the basis of adding a contingency factor to the program schedule. This contingency factor should be stated as such, rather than being hidden in padded estimates of the various tasks. This contingency time becomes the program manager's reserve which allows him to modify the schedule based on feedback, so that the project can be conducted with both flexibility and control.

The estimates of costs to be discussed here are those for the direct costs of the project. Indirect costs, overhead, and general and administrative costs are treated by standard accounting techniques which vary from company to company; however, they are usually assigned as some percentage of the direct costs of the project.

The cost estimate of a project and the operating budget are not the same thing. The cost estimating process should aim at making the very best and most complete estimate of the project's total cost. The estimates should not be padded, nor should they be intentionally low or arbitrarily cut during the estimating process. The operating budget will allocate the minimum reasonable funds to each task, with the understanding that some funds will be held in a strategic reserve controlled by the project manager. It is good management to assign the minimum amount of funds to a task in the operating budget, but bad management to produce a program cost estimate based on the minimum estimates for each task. Here, we are concerned with the preparation of the estimates. Operating budgets are explained later in this chapter. They are not the same thing.

The direct elements of the cost estimate and the operating budget are the same, that is, labor, materials, subcontract costs, equipment, and travel expenses. The estimates of all of these must be honest, because they will be the basis, with the benefit estimate explained below, of deciding whether to conduct the project or not. Perhaps it will be decided to intentionally underestimate the project cost for some political reason. If this is to be done (and it is usually the beginning of a technological tragedy), at least let it be done rationally and consciously, starting from honest estimates. One should try to avoid believing one's own propaganda. Cost estimates generated as unhappy compromises between technical reality and political expediency are one of the biggest traps in real world management of advanced manufacturing technology. Often the reasons for falling into it appear valid. Perhaps only a certain amount of money is available, and if the project is to be approved, the cost estimate must be within that amount. If such is the case, the project should be honestly estimated and divided into phases, with the first phase fitting the available funds. Often this is not done; the program is begun with a budget which is inadequate but misrepresented to be sufficient. When funds run out short of completion, technical problems are then misrepresented to be the cause. Because so much has already been spent, additional funding has to be granted to protect the funds already spent. *Misrepresent* is a management term which means to lie. It is excellent management practice to fire misrepresenters the moment they are discovered.

As with the time estimates, once the cost estimates are prepared, a contingency fund should be added, based on an honest evaluation of the uncertainty of the estimates.

The labor costs of a task are estimated for the various classes of labor—engineering, clerical, shop—depending on how these classes are divided in the particular company. These divisions are usually based, not only on the salary scales of the various skills, but on how overhead and general and administrative costs are assigned. A project manager should make it a point to find just what these rates are, because they can vary substantially. Then, whenever a job is encountered that can be done by either of two skills, the one which is least expensive should be assigned to it. But one must know the overhead rate assigned. It is not uncommon for shop labor to cost more than engineering when this factor is added.

Once the man-hour estimates for each skill on each task are complete, they are converted into cash equivalents. This process is complicated by the necessity to include estimates of how much the various salary and overhead and general and administrative rates will increase during the program. The mechanics of this will vary from company to company and will also depend on the inflationary atmosphere of the time when the project is being planned and conducted.

The cost of materials, subcontract services, and equipment should be based on firm quotations whenever possible. One may not know exactly how many trial pieces it will take to perfect an operation, but one should be able to find out exactly how much each piece will cost. Do not use estimates for any quantity which can be known. Find out.

Travel expenses replace the large overhead expenses associated with company-operated laboratories in projects which are conducted in the applications laboratories of other companies or at universities. They should be realistically estimated, and those estimates defended.

When all of the direct expenses of each task have been estimated or found by investigation, they are summed for the project, and then the contingency factor is estimated and added to it to give the complete cost estimate.

THE PROJECT BENEFIT AND RISK ANALYSIS

Now the objectives of the program, how it will reach those objectives, how long it will take, and how much it will cost are known as well as they can be. One other essential item is required before a decision to go ahead can be made: One must show that the result will be worth the effort and the risk.

Nowhere in American industry is so much misinformation and wishful supposition found as in the supposedly objective evaluation of the potential benefits of new technology and the risks of projects to develop it. This is because the methods of developing and implementing new technology do not control risk. If, at the beginning of the program, no experimental evidence upon which to base the risk assessment is available, because no feasibility study has been done, and if the program plan calls for buying an expensive and untried system based on salesmen's claims and expert opinions, then no honest jutification for the program can be written. But some justification must be contrived to convince management that the program will succeed. Often, these baseless estimates of potential return on investment are concocted to provide some minimum promise of return which company or customer policy requires before the project can be approved. As long as this payback is promised, no one looks very closely at the methods used to produce the promise. In the worst cases, everyone concerned knows that insufficient information is available to estimate the potential payback of the program but agrees to go ahead anyway, using a concocted justification. Advanced manufacturing technology development is essential to corporate survival. However, this does not justify shoveling a company's resources into the wrong projects, poorly planned and managed, without honest justification.

No full-scale development program should ever be approved which does not have a project plan based on a well conducted and documented feasibility study, and in which the expenditure of funds is planned to be proportional to the rate at which insight into the CTQs of the project is acquired. This is the only way risk can be managed. One other important piece of documentation is required, though, and that is an estimate of the value of the project to the company.

If reasonably accurate cost reduction figures can be produced, based on the technical data from the feasibility study, they, of course, should be. The assumptions upon which they are based should be explained in detail, as should their limitations and potential sources of error. If, however, no reasonable basis for quantatively estimating the payback is yet available, none should be concocted. A prose explanation of why such data are not yet available, providing a qualitative (but not subjective) interim justfication for beginning the program, should be given. This explanation must include a discussion of how the necessary data will be acquired, when they will be available, and how much time and money the project will spend before they are available. A management review of the project is then scheduled for that point, and only the resources required to reach it are now approved. Such an alternative is superior to accepting baseless estimates, when it is obvious that no basis for payback estimates can as yet exist, and then losing management control of the program. This often happens when concocted estimates are accepted, because nothing else is available, and then are later found to be wrong. By then, too many conspirators have a stake in ignoring embarrassing evidence of how preposterous the original project justification was to allow its termination.

The estimate of project benefits should be as accurate and as honest as possible. It should generally be more specific and detailed than the one prepared to justify the feasibility study, because obtaining the information to justify the development program is one of the reasons for conducting the feasibility study. The exact format and included information will vary, but it should include all of the things described in Chapter 3 to justify the feasibility study. In particular, the impact of the new technology on product quality must be considered as well as its effect on costs.

The new technology proposed must be compared to something. Vague talk about generalized improvement is not acceptable. The alternatives and why this approach is superior must be explained, as must the risks and how the project plan will control those risks. The implications of not conducting the project should also be discussed. Because all of the promised benefits are dependent on the technology being successfully introduced into production, exactly how and when it will be implemented should also be explained.

An analysis should be included in the proposal which shows that, when the project is complete and the new manufacturing process is implemented, the product will still have a market. This analysis of potential changes outside of the project, which might make the resulting technology irrelevant when finally developed, should also include investigations of other, even more advanced techniques for making the proposed product which might become available.

Sometimes, those working on a project can become so fixated upon their own approach that they cannot conceive of a competitor leapfrogging them. They may even ignore clear evidence that this is happening rather than admit that a good project, carefully done, just took too long to be any longer germane to the company's operations in its market.

Such arguments, however, can also be used as a technique for disapproving any proposal to develop and implement any new technology. Some supposedly even more advanced technology than the one proposed is used as an excuse for postponing any action until it is available and, in the worst cases, has been implemented by some other company. The best way to combat those who will use such excuses for disapproving the project is to address the question forthrightly in the project proposal. Find and investigate the alternatives to the proposed technology and discuss them objectively in this section of the project proposal.

THE PROJECT PROPOSAL

Now all of the information necessary for the development project proposal is available. Because advanced manufacturing technology programs are both important to the company's survival and are expensive, a formal proposal should always be required, no matter what the funding source. This must include the following:

1. An introduction briefly explaining any background information necessary to understanding the proposal, and explaining the project objectives.
2. The final report of the feasibility study.
3. The project task list and any necessary explanaiion, including how the skills of the people assigned will fit the task list.
4. The detailed project plan and any necessary explanation.
5. The project benefit analysis, including consideration of the project's short-term and strategic risks.

Placed in a single binder, this will provide whatever management level must approve of the project with the necessary information to evaluate it. Preparing it will assure that those who will conduct the project understand what they are to do.

Management summaries have become very popular and, if required by the company, can be prepared. My own opinion of such things is that if someone is going to make an important decision and does not have the time or the interest to read all of the available information about that decision, he ought to delegate the job to someone who will.

THE PROJECT OPERATING PLAN

Once the project has been approved, and the resources and schedule are finalized, the project operating plan is prepared. The technical part of this plan is identical with the detailed plan used for preparing the project estimates. Unless major changes have been made in the schedule or resources during the approval cycle, and such changes should be avoided, the technical plan should not change. The logistic support list should also remain unchanged.

However, the financial and schedule estimates assigned to each task should now be replaced with the minimum reasonable time and cost for each item, and the remainder should be assigned to a reserve controlled by the project manager. This minimum figure should be decided by the project manager and the person responsible for the particular task, with the understanding that they will be used to facilitate project control and will be modified as necessary as progress is made.

A plan which takes the best advantage of all of the available information has now been formulated, and the development program can begin with the very best chance of succeeding.

REFERENCES

1. Seelig, W. D., and Rubin, I. M. *The Effects of PERT in R & D Organizations,* published as a working paper of the Research Program on the Management in Science and Technology, No. 230-66. Alfred P. Sloan School of Industrial Management, Massachusetts Institute of Technology, Cambridge, Mass., December 1966. In Moder, Joseph T., Phillips, Cecil R., and Davis, Edward W. (Eds.), *Project Management with CPM, PERT, and Precedence Diagramming,* New York: Van Nostrand Reinhold, 1983.
2. Miller, J. A., and O'Connor, J. J. Inertia and electron beam welded turbine engine power shaft. *Welding Journal,* **59**(5):17–22 (May 1980).
3. Foster R. N., as quoted in Wolff, M. F. Picking the right technology should be the first priority. *Research Management,* 7–8 (July 1981).
4. Mayo, J. Process design as important as product design. *Wall Street Journal,* (Oct. 29, 1984).

6
THE CONDUCT OF ADVANCED MANUFACTURING TECHNOLOGY DEVELOPMENT PROJECTS

INTRODUCTION

There is only one reason for a company to conduct an advanced manufacturing technology project: to improve its ability to compete. To accomplish this, the project must eventually lead to an implemented result in the production shop, otherwise it can have no competitive impact.

There is no technological free lunch. Competitive advantage comes only from the hard work of competent professionals who understand precisely what the company does and how the proposed innovation will fit into and improve the company's operations, and who understand the manufacturing facility in which it must eventually be made to work.

Urgency

In manufacturing technology development, implemented knowledge, leading to a competitive advantage, is the ultimate goal. This is in contrast to basic research, where knowledge is sought for its own sake. This does not make man-tech development less important to mankind or less socially acceptable. It does, however, make a sense of urgency in its conduct much more important. Knowledge is timeless, but the relevance of a specific applied technology is time-dependent. If it takes too long to complete and implement, it might as well not be done.

The idea that the technical quality of a project is directly related to how long it takes is usually wrong. The worst projects are those which drag on while alternating between stupor and frenzy. Because of lack of resources or interest, or because of administrative obstacles, nothing is done between mad flurries just before the annual audit or quarterly report is due. Man-tech programs are best conducted with a relentless Type A drive, a compulsion to get the system built right and up and running well, soon.

Most project delays are logistic and administrative, not technological. A few dedicated engineers and technicians, working as a team at the pace that a difficult and fascinating project can generate, will often produce major breakthroughs in less than a month. Yet, it often takes much more than 30 days to accomplish logistic tasks that an individual, working on his own, would accomplish in a few hours. The requisition, approval, and procurement of the simplest off-the-shelf item can take weeks. Specialized items or services, which do not readily fit into the normal procurement procedures, can take months.

Unfortunately, administrative delay is often considered unavoidable. An engineer who did not perform a technical task assigned to him for several weeks would be in trouble, but administrative or logistic delays are often accepted as if ordained by God. These are the least acceptable of delays, because they result from a procedure or department which is supposed to facilitate the accomplishment of a task impeding it instead.

Manufacturing technology development projects often fit poorly into the normal administrative procedures of companies. Sometimes those who perform those procedures see these projects as being of secondary importance. But more commonly, because the requests of development projects are more work than the routine processing of routine paper, they will be put aside or lost unless constantly pushed by the project staff. It is unfortunate, but the project manager and his staff must realize that to work on advanced manufacturing technology projects is, in many companies, to constantly push, cajole, threaten, beg, and search for lost paperwork.

If this is not done, the project will sink in a swamp of misplaced requisitions, misunderstood justifications, and approvals that can only be signed by someone who is out of town this week. Delays kill projects; they must not be accepted.

Eventually, it is hoped, as companies come to realize the importance of manufacturing technology development and to understand that it is by definition nonroutine, they will provide special administrative procedures in order to expedite their conduct. But the lack of such procedures is no excuse for allowing the existing ones to destroy the momentum and continuity which are characteristic of a well-run program.

Approaches

Advanced manufacturing technology projects can be conducted in one of three ways:

1. By the company's people using the company's own facilities exclusively.

2. By the company's people using outside facilities for some or all of the work.
3. By the company's people working with outside vendors.

Projects conducted in-house range from small improvements of existing functions to major development efforts. Which development efforts are done by this method will depend on the perceived relationship between the company's products and the processes which make them, and a definition of just what is the company's business.

Deciding whether to buy the components of a system or subsystem separately and assemble it in one's own facility, or to develop it with a subcontractor and have him supply it complete, must be answered by weighing a number of factors. The scheduling, the cost, and the need to protect proprietary information must be weighed against the question of the company's mission, limited resources, and the inescapable increase in risk that results from connecting pieces obtained from various sources together into an integrated system. The principle that both structures and projects most often fail at the joints must be constantly remembered.

A rational decision that weighs the various factors of the specific project should be made in selecting the approach. Nothing should be allowed to happen because it has always been done that way, or because it is decided by not making a decision. It is surprising how many times things are done in ways that cannot survive reasoned consideration because no reasoned consideration is given to them. These decisions-by-not-deciding come about because everyone accepts a given approach without thinking about the alternatives. Failures to realize that alternatives exist and consciously select the best from among them are amazingly common. It can be avoided by constantly asking, why are we doing it this way? Is there a better way? Sometimes the answers can be very surprising.

Honesty

The conduct of advanced manufacturing technology projects is the sequential solution of problems. No problem can be solved if its existence is not admitted, or if no working solution has been truly found but is misrepresented to have been found. Projects have failed not because of any basic technological impossibility, but because positive interim results were fabricated, or because evidence of unsolved problems was suppressed. This dishonesty is always rationalized as being for the good of the project or the good of the company. If a project can only be approved, or continued, if those who evaluate it are misled about it, then that project should be immediately terminated. But what if those doing the evaluation want to be misled?

Too often the corporate culture and unwillingness to admit and manage the inherent risk of man-tech development creates an atmosphere where distortion of results is not only accepted but almost required. Cognitive dissonance is the "psychological conflict resulting from incongruous beliefs and attitudes held simultaneously."[1] It is the tendency "to suppress, gloss over, water down, or 'waffle' issues which would produce conflict or 'psychological pain' within the organization."[2] The project manager and his staff have produced a technical plan, a schedule, and a budget based upon sound but necessarily limited detailed knowledge of the new manufacturing process under consideration. During the conduct of the program to the plan, it is found that technical reality requires that, although the approach being developed will work, it will take longer or cost more than planned, or perhaps the technical approach, approved by management, must be modified.

What does the project manager do? In the dream world of management literature, he and his crew decide the best way to deal with the problem, put together a logical explanation for a benign management who listens, nods knowingly, and provides the necessary resources to solve the problem so that progress can continue.

In the real world, cognitive dissonance is too often the answer. The unfavorable results are suppressed, ignored, or edited, and what might be a relatively easily solved problem is allowed to fester into a project-killing calamity.

American management, in the last 20 or 30 years, has jumped from fad to fad. Too often, sound concepts have been misapplied by those who were looking for cheap, quick, and, above all, easy routes to corporate success, without having to understand the basic principles upon which the concept was based and without providing the necessary prerequisites for it to work. The introduction of management by objectives into crisis-run organizations where no plan is followed for more than a week, and the introduction of quality circles into companies geared to wringing every last drop of short-term profit out of every shoddy product shoved out the door, are two examples.

Now, when the necessity for manufacturing companies to concentrate on improving their manufacturing technology is finally being recognized (and the fact that it has been ignored by many companies for many years is in itself an example of cognitive dissonance), we have the "project champion" riding onto the scene. Studies of successful projects show that they succeeded because someone overcame all of the obstacles which companies erect to protect themselves from acquiring the competitive advantages which will help them survive. The project champion is supposed to have the technical insight of Tom Edison, the cunning of Sergeant Bilko, and the ambition of Attila the Hun. As long as a company has a good supply of such people, the organizational problems which make them necessary to project success can continue to be ignored. Again, the fad is attractive because it promises an easy way around problems that management does not want to confront. One does not have to provide the people, facilities, or funds for man-tech development, or find a way to reconcile the requirements of the new technology with the old labor agreement, or even make an honest and complete commitment to technical improvement. All that management must do is find a project champion, and he will create success—quick, cheap, and easy.

Yes, nothing gets done unless someone makes sure it gets done, and the improvement of manufacturing technology is, even in the best of companies, difficult. The project champion is really just an honest, competent project manager doing a first-rate job. The problem is that the project champion, especially in a third-rate, fad-afflicted company, can and sometimes must become the project pimp.

To push and to protect the project forcefully, yet remain brutally honest about it with oneself and with a management which doesn't want the truth, the whole truth, and nothing but the truth, is not easy. But it is necessary to the final success of the project. When the system the project develops eventually gets to the production floor, any unpleasant truths are going to be obvious to everyone, so problems had best be attacked when first detected.

The recuperator hole periphery welder (Chapter 4) is an example of a project that only succeeded because its technical problems were honestly admitted and eventually solved. The project was sponsored by a U.S. government agency, and all of the difficulties were honestly and promptly reported to them. They, in turn, behaved in a professional manner and supported the efforts to solve those problems. It is not surprising, however, that someone is less than understanding if he is fed propaganda and only told the truth when it has become so monstrous that it can no longer be suppressed.

The development of insight about the technology by those who manage the project, and honesty in evaluating results and controlling the project, are absolutely required for its success. No system of planning, scheduling, or auditing can save a project if these essential elements are not present.

In-house Projects

All advanced manufacturing projects should be in-house projects even if most or all of the development work is conducted in subcontractor or university labs and shops. These projects must be conceived, planned, and conducted by the employees of the company whose manufacturing facility will use the technology to make its products. The company's people must implement the system in production and later provide technical support to it. But those portions of the technical effort of the project which are to be performed on the company's own premises present some different management problems from those projects or parts of projects done in subcontractors' facilities.

There is a problem of priorities and interruptions for those doing the work. The knowledgeable, dedicated people who will do the best development work are usually those who are most in demand to troubleshoot problems on the existing technology and provide consulting services in their specialties to other departments. These interruptions, each of them justified, can make development work impossible. To really dig into a problem, and eventually find not only a solution but the best solution for the given application, not only takes time, but takes time in large, unbroken blocks. In order to provide this freedom from interruption, policies which isolate development activities from other operations are required. No policy, though, will keep a desperate production supervisor from tracking down the person he knows will get his stopped production line running again. But one must try. Labs should be physically isolated from production for this reason, and some sort of buffer placed in the way to assure that only the most important interruptions are permitted. This is another critically important function of the Secretary. All requests should go through her, and she should be knowledgeable enough to be able to sort them by true urgency and crafty enough to deal successfully with those making the requests. This is not an easy job, and it is very important.

There are other reasons for isolating the location of the conduct of advanced manufacturing technology development from the manufacturing shop. Once a working, debugged system or process is completed and tested, the most difficult phase, production implementation, begins. A key to implementation is convincing the production shop people that the system really works well. No system works well during the development phase. If it did, it would not be undergoing development. There is an old New England proverb which states that one should never show a fool a half-finished job, which, while perhaps too critical of manufacturing personnel, conveys the essential point. One should never go to the production floor with a new system until it is completely ready, and one should be very careful about which production people are brought into the project and when. When production people are introduced to the system, it should be with a professional presentation and demonstration on the project, and a serious dialogue. But manufacturing acceptance of the new system is definitely not facilitated by allowing everyone in the shop to casually wander in to scoff.

Another problem which is frequently encountered in in-house development programs is being constrained by the available equipment. If one's efforts are focused onto originally picking the best technology and developing it to meet the company's needs, rather than using a different approach for every project, such constraints can be an advantage. If one is forced, however, to always use the same technology, as it grows steadily more obsolete and regardless of its suitability for a given application, then obviously it is a disadvantage. Unfortunately, these two situations are not alternatives, but the phases through which most man-tech development facilities pass. Very few are ever disposed of when they should be. This is because both production and development manufacturing equipment is usually assessed for replacement by the same standards. It is quite difficult to explain to a nontechnical person, particularly from a production shop background, that a machine that may have many years of useful shop life left is obsolete as a tool for advanced technology development. Often it is easier to replace the one in the shop, because one can show financial return on investment and the production of money is easier to explain and justify than the production of knowledge. This can lead to the odd situation where obsolete production equipment, or new equipment found to be unsuitable or undependable in production, is moved into the advanced manufacturing technology lab. When this is suggested, one should carefully explain the difference between a development lab and a museum to those making the suggestion.

Certain types of development projects are most suitable for in-house labs. These are projects where none of the central technical questions involve the equipment itself but rather are concerned with the application of the equipment to a specific operation. For example, one might first conduct a program to select a specific type of machine to perform a generic operation, such as electron beam welding. A given high- or low-voltage system might be selected and the techniques for the first production applications developed. A laboratory machine to develop follow-on applications or new products is an asset, even if one or more similar machines are operating in production. Borrowing time for development work on production equipment never works very well. The difficult project with the biggest potential payoff is the one most handicapped by the need to constantly step aside to allow production use of the equipment.

But, as time passes, the lab machine, which may be the first one bought, begins to prevent the exploitation of newer developments because it does not have the required features, and because it is there. This problem can be aggravated by the very dedication of lab people and the way they often care for their equipment. It is quite common for the old lab machine to be in far better condition than newer production shop equipment which was maintained by the "run it till it stops with a bang" method. Because of this, the still shiny old lab machine is not considered for replacement or overhaul and retrofit.

When projects are done with such old equipment, there may be considerable resistance to trying the newer approach or even mentioning it to management. This may be caused by the same fear of change that lab people hate so in production personnel. But it also results from the feeling that one is expected by management to do the assigned job with the assigned facilities, and it is better to come up with some result, even if it is unduly complex, than to be disruptive by requesting the new facilities which would permit more elegant solutions.

Another problem of in-house development labs is that they keep one in-house. By providing a single type of facility in a single place, one loses the opportunity to investigate many types of facilities and know at any given time what the best available technology for an application is.

All of this results from failing to recognize that the in-house capability to develop and implement advanced manufacturing technology is not having in-house machinery, it is having employees with the required skills. Competent people are the only essential facility. That few companies realize this is obvious from how they attempt to impress visiting corporate officers and customers with their capabilities; they take them on tours and show them equipment. If they understood the nature of this capability, they would instead have their best people give presentations on their work.

When conducting advanced manufacturing technology development projects using in-house facilities, one must avoid having a single preconceived solution to a problem which must be made to work no matter how poorly and at what cost. The manager must assure that using the same equipment continually does not cement his people's minds closed to other, possibly better, approaches.

A DOCU-FICTION CASE STUDY: THE ROBOTIC X-DEVICE BENDER

In the preparation of this book, I talked to many people who told stories of projects which, while costing fortunes, led nowhere. Many seemed to follow the same scenario: All failed to produce a production-implemented technical improvement because, while no one seemed to have done anything really wrong, what was done right was somehow not done right well enough. To avoid embarrassing any person or company, I have combined the important common elements into the following generalized case study.

A well-established company manufactured a line of related products, all of which used an X-device as a central component. This product line and the company had grown because they had been the first in their industry, 20 years before, to invent and find a way to make a bent X-device. It was technically superior and significantly less expensive than the welded type which it replaced in their products.

Some of their competitors had continued to use the welded component, improving it and cleverly designing their products so that its inherent disadvantages were compensated for by other components. Other competitors had adapted the bent X-device when its advantages had become obvious and, in the last few years, had been steadily reducing the original company's lead and market share.

In the previous year or so, both the trade journals and the general business press had been running articles on robotic X-device benders, which promised to produce not only a cheaper product but a better-quality one. Also, the Japanese were supposedly about to target the industry. Even more worrisome to the company's management were rumours that their closest competitor was about to perfect a working twisted X-device which was technically superior to the bent one and could only be made with robotic equipment.

Top management therefore decided that they must do something about the robotic problem. It would not go away. Whether they were motivated by a genuine concern for the future of the company or were worried about their own image is not important to our story. They looked around the company, and because they had no robots, decided that therefore no one working for them could know anything about robots. Just to be sure, though, they asked the manufacturing manager if he thought that any of his people might be able to manage a major robotic project. There were two employees he quickly considered and rejected. One was the production department foreman, who was smart and dedicated, knew the X-device business completely, and had been with the company for 19 years. He did not have a college degree. The other choice was a manufacturing engineer who did have a college degree and who, in the six years since joining the company, had made a number of significant improvements to the existing X-device manufacturing process. He was also the fastest troubleshooter they had ever had, with a talent for looking into a complex electrical-mechanical problem and instantly seeing the key to its solution. The manufacturing manager decided that this insight into the old technology might prevent him from being open-minded and innovative about the new.

Besides, the manufacturing manager needed him. Faced with the constant need to meet production schedules and budgets, he really did not believe that a new system, still several years away, should be given priority. He had become the manufacturing manager because of his skill at meeting the short-term goals by which the manufacturing department, of necessity, operated. He needed both the foreman and the manufacturing engineer to meet those goals. He did not intend to disqualify them for this new job because of their dedication and competence, although that was the result. He honestly felt that his needs were more important to the company's success than the development of some far-off technology which might or might not prove beneficial.

A person was hired who appeared to the personnel department interviewer who chose him to know a lot about robots. Of course, the interviewer did not know anything about robots, so his ability to judge the applicant's depth of knowledge was limited. The person selected came from a different industry and therefore knew practically nothing about the company's business in general, and absolutely nothing about X-device manufacturing in particular. Actually, until the personnel man took him on a tour of the plant toward the end of the interview, he had not been aware of the existence of X-devices.

Unfortunately, this total ignorance was not obvious to the interviewer because he knew very little himself. However, it was obvious to the supervisor and workers in the X-device manufacturing department, who gave this new technical employee the respect they always extended to new white shirts without insight about the product or processes: "Oh *blank*, another one of those dumb *blanks* come to *blank* up the *blank*ing operation."

All this notwithstanding, he was hired. The personnel department had been given the task of finding a robot expert and they had accomplished it. He was given the mission of "getting us into the robotic X-device business" by upper management. They, too, had now accomplished their task. They had defined the problem in general terms and delegated it to someone they were assured was capable of solving it.

The newly appointed project manager now prepared to tour all equipment manufacturers who advertised working robotic X-device benders for sale. He first went to the library and read the cursory and often contradictory articles he could find on their manufacture. There was practically nothing on the old technology, but there were the articles on the new which had originally prodded upper management into action. These, when read by someone desperate for hard information, were found to be contradictory and incomplete in many details.

Now, with at least superficial information about the company's business, he returned to the production shop and introduced himself to the supervisor again, explained his new job, and was, in turn, introduced to the manufacturing engineer who was in the midst of troubleshooting a complicated, intermittent equipment problem. If anything, he got a worse reception than on his first visit. He had acquired enough partial information and buzz words from the literature to ask really dumb questions. His reception was not improved by both the supervisor's and the engineer's thinking that he had a much better job than theirs, one for which they had not even been considered before it was awarded to this ignoramus from the outside.

The project manager's self-esteem quickly recovered, however, once he was on the road. The equipment manufacturers' sales and technical personnel treated him like the greatest living expert on X-device manufacture. This was genuine, in some cases. After all, he did hold an important technical position with one of the leading manufacturers of X-device-based products. If he seemed a little superficial on the actual details of the processing of this component, it was obvious to only a few of the technical people who had worked in the industry before joining the equipment manufacturer. In most cases he did not spend enough time with them to be found out. The sales people with whom he mostly dealt did not have any more technical insight about the business than he did.

Upon his return, he prepared a long report of what he had seen, enhanced by information on sizes and speeds from the various pamphlets he had collected, and carefully comparing the various vendors' machines. He also prepared a one-page executive summary which his secretary flawlessly typed on a word processor. This machine permitted her to print all of the key buzz words more prominently that the rest of the text. Although no one realized it at the time, the typing of this summary was to be the supreme technical accomplishment of the entire project.

Top management, honestly trying to avoid micromanaging the program, and impressed by the apparent professionalism of the report, gave the project manager permission to proceed with obtaining detailed proposals for a system. The memo which did this was addressed to the purchasing department manager and asked that he provide all possible assistance in the solicitation of bids and the preparation of the capital equipment proposal.

The manufacturing manager was sent an information copy of this memo with the full report attached. He passed it down to the production department supervisor and to the manufacturing engineer, in order to assure vertical communication and inform them of the long-term plans for their area. This was supposed to assure their cooperation.

They went through the report and found several important questions about the application of the proposed technology to the company's operation which the project manager had not asked because he was not aware they existed. They also noticed that several of the factors which were supposed to be either essential or important advantages of the equipment he recommended were irrelevant, overly complex for their operation, or technically trivial.

When they raised these points with the manufacturing manager, they were told not to be concerned. The project manager was an expert and would solve all those problems, which they were probably exaggerating anyway.

All of this communication had an interesting effect on the supervisor and the engineer, the two people who would eventually be most important to the successful implementation of the robotic X-device bender. On the very first day of the project they were praying for its failure. These two, and their people, had steadily and incrementally improved both the existing process and the resulting product, and they felt that the company would not have maintained its position in the industry without their efforts. They also felt that top management was obviously unaware of their efforts or contemptuous of them from the way the project had been conducted to date.

This did not improve their day-to-day performance either, although they were so committed to the production of the perfect X-device after years of striving for it that even with their morale destroyed, their performance was still excellent.

The project manager now constructed a set of criteria the new system must meet. These included such things as accuracy and precision, ease of programming, safety, type of memory, and production rate—the things that manufacturers put into their brochures. It did not specifically mention the company's most important X-device, the one used in its newest and largest product line. This was a double-filleted design which required a wedged back bend. Although there were eight vendors offering robotic X-device benders, no one had ever even attempted to make a double-filleted one robotically. No one else in the business knew how to make it by the old technology, either. So the potential vendors did not realize how difficult the problem was or how elegant were the wedged back bending techniques which the manufacturing engineer had devised.

Requests for quotation were now sent to all of the equipment manufacturers, soliciting specific information on each of the factors the project manager knew were important and, of course, not mentioning the important factors of which he was unaware.

He did not want to select a system based only on vendors' paper promises, however. Therefore, the request for quotation required that each vendor submit six standard X-device specimens for evaluation with their proposal. Because the raw blanks from which X-devices were made were quite expensive, a test specimen similar to a low-cost generic part had been developed for qualification of equipment and processes. The specimen had been in use for many years throughout the industry, and as part designs had evolved and become more sophisticated, the specimen had become less and less representative of actual parts. Those who were not current in the technology of the product and familiar with the specimen did not realize how far this divergence had progressed, and that in order to overcome it, different classes of specimens had been developed to better simulate specific types of parts.

Detailed discussions were held with each potential supplier about the system he would quote. These were relaxed, amicable meetings because the project manager and the vendors had essentially the same view of the technical questions involved. This was because the project manager's perspective had been gained primarily from his discussions with the vendors and his study of their literature. He did not have an understanding of the technology from the viewpoint of how it must be applied to his own company's specific products and how it would mesh with its existing processes and facilities. If he had possessed this divergent view from the suppliers and had had more discernment about the problems he faced, these meetings would have been considerably more intense. Rather than merely reviewing and ratifying a set of subconsciously preagreed concepts, it would have been necessary to define and reconcile a number of quite different views of the problems the project would face. Successful joint customer-vendor man-tech projects generally begin with intense meetings requiring considerable preparation and concentration on both sides, and end with relaxed luncheon meetings in good restaurants. Failures usually start with multi-martini lunches and end with desperately intense meetings.

The project manager certainly did not intend to fail or underrepresent his company. He made sure that each potential supplier knew that his proposal would be evaluated against those of his competitors based on promised production rate, system cost, and delivery schedule; on the quality of his submitted specimens; and on the proposed terms of his warrantee.

Two of the potential suppliers did not quote. One did not quote because the sales manager left the company suddenly, causing such confusion that the filing date was missed. The other, who chose not to quote, was one of the newer companies in the field but one which had gained a reputation for technical excellence and aggressive competition. This made their refusal surprising, but they excused it by saying that they were fully booked with orders and did not have the resources to become involved in such a large and important project at this time.

As far as it went, this was true. They were booked, and their president was concerned with keeping the company's growth to a controllable and sustainable rate, which he felt was limited by their ability to hire first-rate professionals and absorb them into the organization. While negotiating with this potential customer, they had also been asked to bid a similar system by one of its competitors. They chose to bid this other job because they could not do both and they were somehow disturbed by the technical superficiality with which the discussions were conducted, compared to the other company whose representatives really knew what they wanted and needed.

Six proposals were received and reviewed by the project manager and the manager of the purchasing department. The specimens each potential supplier submitted were sent to the quality control lab for evaluation. Four sets met the specification requirements. One set was made from a nonstandard material which happened to be all the vendor had available. The other set, for some reason, had only four rather than the required six pieces.

Of the four bidders who had submitted the required specimens, one was eliminated because investigation showed that he had never built a complete system of the size and capacity required, and another was eliminated because his price was 30 percent higher than the next-highest bid.

There was a problem in deciding between the two remaining bidders. The system quoted at the lowest price, best delivery, and best warranty terms did not have several of the features that the project manager thought important. The purchasing manager had called the past customers of both companies and had found that the low bidder had previously supplied equipment which had not been successfully implemented.

The next-lowest bidder, whom both the purchasing manager and the project manager preferred, was therefore contacted and told that he could have the order if he could improve his delivery date and warranty conditions. In particular, he would have to guarantee a substantially higher production rate than he had originally quoted. The two managers did not force him to meet the price of the lower bidder, because they both felt that it would have been unethical to inform him of the exact number. They did, however, ask that he review his pricing with a very sharp pencil. The project manager honestly felt that this vendor had offered the best overall deal for the company, but that in order to get top management's approval for this selection, the proposal must be superior to the low bidder's in all factors.

The vendor, although his original bid had been very tightly calculated, and his promised production rate, in particular, was the very maximum his technical people thought realistic, sharpened his pencil and complied, because his management wanted so badly to win. Receiving this order from one of the top X-device producers was viewed as a major competitive accomplishment. It would increase the company's chances of winning several others for which they were also bidding.

A presentation, based on this final submittal, was now made by the project manager to top management. He impressed them with the technical promise of the proposed robotic X-device bender. They were pleased that the proposed price was close to their original estimates of what such a facility would cost. The Gantt chart showing the schedule for the project seemed reasonable, especially if the substantial progress the project manager had already made in his first few months on the job was considered. He was congratulated on his excellent performance and given permission to proceed with the project.

The purchase order was issued and work on the project formally began. The vendor's chief engineer realized the tight budget and time constraints he must operate under and knew that the system must be a success because of its high visibility throughout the industry; but more important, it had to succeed because it was his system and he had great pride in his work.

Preliminary design work had actually started before receipt of the written purchase order because certain long lead time components had to be ordered early if the system was to be delivered on time. Now, when detailed design of the system began, certain questions were encountered about the specific application of the equipment. The vendor's chief engineer had, of course, visited the customer's shop during the original prebid process and had been given a brief general explanation of the process by the project manager. The questions which now emerged were much more detailed than the information originally provided could answer. The project manager was contacted with a list of these specific questions, and he returned a set of answers based on the engineering drawings, manufacturing operation instruction sheets, and building and facility drawings. He included copies of these documents.

This information, when closely studied, raised more disquieting questions than those it answered. There were a number of areas where the documentation subtly contradicted impressions about the customer's procedures derived from conversations with the project manager. Certain assumptions about the finished X-devices and the requirements the customer's overall product design placed on them had not been clearly communicated because the project manager had not completely understood these factors himself.

The chief engineer called a meeting of his sales manager and company president and explained his misgivings. In particular, he needed detailed information about the wedged back bend on the double-filleted X-device. If this component was important to the customer's business, certain of the already ordered long lead time parts of the new system might have to be changed. Perhaps it was just a misunderstanding, but it had best be addressed quickly because of the potential impact upon the project if the system design had to be changed.

It was decided that, to avoid alarming the customer, the sales manager would visit his plant and try to clear up what they all hoped was simply a misunderstanding. In the meantime, the chief engineer would continue with the design based on the original assumptions.

The sales manager met with the project manager and, after a fine lunch, raised the questions the chief engineer had asked in a way that did not convey the chief engineer's degree of concern. The sales manager's professional life had been about the worst possible training for grilling a customer about issues the customer might not be able or willing to explain. He was also not very good at telling customers things they might not want to hear. He was, however, rather charming. So the questions were partially asked and partially answered, and everybody stayed happy. For a while.

The sales manager reported back that as far as he could see, the chief engineer's worries were exaggerated, and there was no reason to get involved with a big redesign. The answer he brought back to the key question about the double filleted X-device was that it didn't seem all that important, as long as a reasonable production rate and quality level were maintained. The word *reasonable* was left comfortably undefined.

Design and then assembly of the system progressed, the first milestones on the Gantt chart were met on time, drawings were prepared, and parts orders were placed. However, the first major milestone slipped by almost six weeks because of the late component deliveries. Whether these components were actually late or whether the schedule had been based on overly optimistic estimates was never established.

The project manager was far too competent to allow the vendor to put the entire system together without having a required demonstration when the first major subsystem was ready to run. A progress payment was dependent on this demonstration. Because of the late deliveries of components, not only was the assembly late, but very little of the planned time for trials and debugging remained. The vendor actually got the subsystem running, after a fashion, on the afternoon of the day before the twice-postponed demonstration was to take place.

The project manager was fortunate in one way though; the manufacturing manager, who was originally scheduled to accompany him, now had a conflict, and the purchasing manager accompanied him on this, his first trip to the vendor's factory since his original tour. In order to save both time and travel costs, they caught a dawn flight and arrived at the vendor's shop at 11 o'clock in the morning.

After a briefing on the major successes and minor problems so far encountered, and coffee served by the president's elegant secretary, they strolled into the shop for the demonstration. First, the subsystem was explained and its best features pointed out, and then it was run. It had completed two operating cycles when, in the middle of the third, it quit cold. After some embarrassed joking by the president and sales manager and some frenzied activity by the engineering staff, the chief engineer announced that it was just a minor electrical component in a motor driver which they would change in a few minutes, as soon as the electronics technician returned from lunch.

He did not mention that this was the twenty-fourth "minor" failure of a part in the main motor drive unit in the seven, round-the-clock, days since they had first tried to run the machine, or that he was just about sure that major parts of the drive system were insufficient for the loads placed upon them. These minor electrical components, he was afraid, were just the weakest links in a too-flimsy chain.

The sales manager suggested that they all "grab a quick bite" while the component was being changed. The place he chose for this quick bite was a gorgeous expense-account restaurant 20 miles away. He and the president presided over a fine meal, beautifully served. This not only kept them out of the shop until 2:30, but created an atmosphere of mellow gratitude which was helpful when the same component blew on the first cycle of the next attempt to demonstrate the machine.

By then it was necessary to leave to catch the return flight. As he was driving them to the airport, the sales manager explained that silly problems like this happened with machinery all the time. It always glitched at the least convenient moment. Besides, they had seen the system operate successfully several times, and if it could cycle three times, it could be made to work flawlessly. The project and purchasing managers both had reasons other than the residual wine from lunch for accepting this non sequitur; they had selected the vendor.

Upon their return home, the project manager wrote a trip report with which the purchasing manager concurred. The key portion of it said:

Although the project has fallen slightly behind schedule due to the late delivery of an important part from a subcontractor, the vendor has met this important milestone. The key subsystem was successfully demonstrated. Its functioning met essentially all requirements. Unfortunately, the failure of a minor electrical component precluded the measuring of the production rate in a full one-hour trial, but a calculation of this rate based on the measured cycle time during operation shows that the production rate will meet or be very close to the guaranteed value. The system was tried using standard X-device test specimens, and three of these were brought back and submitted to the Q.C. Lab for evaluation. They appear to be of good quality. The results of the lab evaluation will be reported as soon as available.

One specimen was found to be barely acceptable, one was a borderline reject, and the third was found to have a very bad defect. It was also noted on the lab report that these were class I specimens, and the company normally used class III specimens in the testing of its process equipment because they were more representative of its product.

The project manager was confused by the results, and when he asked the laboratory supervisor about the implications, did not become less confused. He was told that a machine might make good specimens yet not make acceptable parts. The converse was sometimes also true. Because these specimens were of an inappropriate class, which meant that they were of a material very different from that which the company used, no conclusions or predictions about the machine's performance could be made. The project manager decided not to report these results after all, because they appeared to be worthless. However, he did not retract the approval which had been based on the specimens.

Meanwhile, the first signs of cognitive dissonance appeared in the vendor's organization as well. Immediately after the project and purchasing managers had left for the airport, the chief engineer and his electrical people sat down to discuss the failures they had encountered that day. They were faced with two approaches to the problem: They could continue to replace each component, as it blew, with one of higher capacity, until either the failures stopped or the entire system had been redesigned one part at a time. There were thousands of parts. Or they could admit that all of the supposed minor failures, taken together, constituted a major problem and solve it by replacing the entire motor system. The consensus of the technical staff was that the second approach was the only sensible one. The soldering iron hadn't been unplugged in a week.

The chief engineer went to see the president and explained the problem. He also explained the implications of the motor drive system replacement; it would involve replacing all of the motors, and the motor controller as well. This, in turn, would cause a problem in interfacing the new system with the existing computer. It was going to cost a substantial amount of money, and it would mean several months of delay while the new parts were obtained and installed. Similar changes would be required in the other major subsystem, assembly of which was about to begin.

The president asked a couple of clarifying questions, then pointed out that perhaps the part which had blown that afternoon was the last weak link. It would be unfortunate if they now replaced the whole motor system and jeopardized the contract, after they had actually solved the problem. He then explained that because they were behind schedule, because the price they had been forced to quote was so low, and because the customer had approved the system that afternoon so they could get a progress payment, he really did not want to start redesigning and rebuilding it now. The reasons for avoiding the problem far outweighed those of attacking it frontally. When he left the president's office, the chief engineer accepted the necessity to continue patching up what the technicians now referred to as "the Kludge."

The project manager, although he let the impression created by his trip report stand, did attempt to clarify the specimen results. He called the president of the subcontracting company and asked if it would be possible to repeat the test with class III specimens. He also asked if the motor driver was now working dependably.

The president explained that after replacing the component which had failed on the afternoon of the demonstration, there had been no more failures. He then talked about infant mortality in electronic components; early in system start-up a few weak parts always fail, but those that survive the first few hours of operation usually last for years. They did not expect any more problems. He did not mention that the system had been only dry cycled, with no parts or specimens being processed, since the replacement of the last component. He apologized that it would not be possible to repeat the trial, because the machine was now being integrated into the full system. If the customer wanted them to stop this assembly and set the machine up for another trial, of course, the vendor would do it. But it would involve some additional expense for which he would have to ask payment. It would also result in several weeks' delay.

They decided that it made more sense to go ahead toward the trial of the full system, the next major milestone, than to go back and repeat the semicompleted one. The president also assured the project manager that in this case the differences between class I and class III specimens were inconsequential.

While the building of the complete system proceeded, the project manager began to make plans for its installation. He had drawings made to rearrange the production department to provide space for it and to provide the necessary utilities. The production supervisor was consulted and he, in turn, consulted the manufacturing engineer. They had heard about the specimen results from the laboratory supervisor, and they now went to the manufacturing manager and explained that from what they could find out, this totally untested system was to be dumped into the middle of the production department. Worse, before it even arrived they would be shut down for several weeks to do the necessary rearrangement.

The manufacturing manager saw himself trapped. If he did not allow the new system to be installed, he would be blamed for its failure. If he allowed the system to be brought in, as planned, his operation would be badly disrupted. The manufacturing engineer advised that before accepting the system and the resulting disruption, someone who knew something about what the system must do should look at it and assess its chances of working.

The manufacturing manager recommended this to top management, both because it seemed best and to get himself out of the box. However, from what he heard about the specimen results, he was afraid that such an audit might be very negative, especially if he allowed the supervisor or manufacturing engineer to survey the system. The comments they were making about the project and its manager made that very clear. He knew that if they talked this way before top management, they would be seen as antitechnical reactionaries, and their honest concerns would be ignored. He told them he would take care of it and thanked them for their concern.

He saw a way out of the problem. Rather than send one of these two fire breathers out to check on the equipment, he asked the manufacturing engineer's boss to do it. This gentleman, while never having been personally involved in X-device manufacture, at least knew something about it, and he seemed to have no prejudices about the project manager or how the project had been conducted.

The senior manufacturing engineer met with the manufacturing engineer and production supervisor, who gave him a view of the project which was based on the few written reports which they had seen and contempt for the professional competence of the project manager. This latter factor appeared to be the classic lack-of-communication syndrome. The project manager had irritated them from the beginning. His reports, not so much because of the answers they gave as because of the questions they never raised, had turned this irritation into contempt for him and for his project.

The senior manufacturing engineer finally managed to cut through this wall of prejudice to determine their concerns about the system. This was, quite simply, that the system would not dependably run production parts. He understood this concern about the production suitability of the system but really did not think that only they knew how it should be done. After all, both the project manager and the equipment vendor's staff were acknowledged professionals in the field. It really did not seem plausible that these two men, despite their skills with the old equipment, should have much to contribute to a new, far advanced technology.

However, their fears about the production suitability of the new system were exactly those of the manufacturing manager, so he accepted their suggestion that rather than watch the system only run specimens, the senior manufacturing engineer should take a dozen double-filleted X-device blanks with him for a trial run.

The project manager, when this suggestion was raised, pointed out that the project plan called for the system to first be tested on specimens, then for parts to be run. However, because the program was now several months behind schedule, he had no objections to doing both on the same visit. When he told the vendor of this plan, the vendor agreed, with the exception that the progress payment due after demonstration of the complete system on specimens would be made, even if problems were encountered with the parts.

The trial was twice postponed because of delays in the integration and debugging of the complete system. It was now six months behind schedule. Again, the meeting started with coffee served by the president's elegant secretary, but the air of amicability of the previous meeting was now replaced with tenseness. The vendor had questions about the box of X-device blanks which had arrived two days previously for the test.

These would not fit into the system's input hopper. When one of the new blanks was placed on the table next to one of those given to the vendor a year and a half before, at the beginning of the project, it was found to have significant differences. The original blank was the older single-filleted type which the project manager had supplied because at that time he did not know the difference. In addition, the new blank had an additional flange which was shown on none of the supplied engineering drawings.

A phone call to the customer's design department found that the flange had been added eight months before, when production of a new product model had started. Because neither the project manager nor the vendor were on the routine distribution for engineering drawing changes, neither had been informed. This discovery did not reduce the tension in the room. The project manager postponed their return flight by 24 hours and reserved rooms at a nearby hotel. It seemed that each question they tried to answer raised two or three worse ones.

It was decided to test the system first by running the class I specimens the vendor had used during his setup and debugging. A load of the class III specimen blanks the customer routinely used would then be run.

While this was being done, the vendor's engineering staff would see if it was possible to hand feed the part blanks with the input hopper removed. This would determine what system components in addition to the input hopper needed modification.

The first part of the test went well; 25 class I specimens were run without a malfunction. The hopper was then loaded with class III specimens. As the second blank entered the system, an electronic component blew out in the motor controller unit. While they waited for it to be repaired, the president talked about infant mortality of electronic parts; the weak ones blew when the system was first tried, but those that survived the first few cycles would probably last forever. He seemed to be telling this story a lot lately.

The chief engineer was paged just then and excused himself. As the repair was completed, he returned to ask the president if he might speak to him for a minute. As they left the shop, the system was restarted by a technician who, when it immediately blew again, commented that they could change parts forever, but the motor system would never be powerful enough to run class III specimens.

The project manager and the senior manufacturing engineer hurried back to the conference room where they had left their briefcases. The time had now come to read, really read, the purchase order and specification for the machine. Presently they were joined by the vendor's president, sales manager, and chief engineer. The door was closed, and the bloodbath began.

First the president announced that running the part blanks would involve a great deal more than just changing the input hopper. Significant redesign of the system would be required. The cost of all of this work was obviously the customer's responsibility since he had changed the part design without informing the vendor.

The project manager replied that although the drawing change did cause the hopper to need modification, in no way could all of the other changes be blamed on the additional flange. Besides, the system could not even run class III specimens. How did the vendor ever expect it to run parts? There was a written warranty of the system's performance, and until it worked to the specification requirements, the vendor had no chance of collecting even the agreed price, let alone any additional costs.

The sales manager intervened, suggesting that rather than shouting at each other, they should try to rationally agree on what the problems were, and the causes, before trying to decide on solutions. He went to the blackboard and wrote "Problem # 1" in the upper left-hand corner. Then everyone resumed shouting at each other.

Eventually the problems were defined as follows:

Problem # 1. The specification for the system required that it be tested first by running "standard X-device test specimens." It did not say what class of specimens. The vendor contended that this meant class I specimens and that if the customer had intended some other type to be used, he should have specified the type, or at least have informed the vendor at the time of the subsystem demonstration and acceptance with class I specimens. Therefore, the costs of installing a more powerful motor system should be borne by the customer.

The customer contended that because no specimen class was given, the system should be capable of running all classes. Besides, if the motor system the vendor had selected was so marginal that it could only run class I specimens, it was obviously unsuited to the intended purpose in a production environment.

Problem # 2. The system would require considerable redesign and rebuilding before it would be able to make the current production model, double-filleted, back bent, X-device. The customer contended that the manufacture of X-devices, not specimens, was obviously why he had ordered the system. He had provided the vendor with all of the information he requested. It was obviously the vendor's responsibility to build a system which could run the customer's parts. That was, after all, what he was being paid to do.

The vendor replied that the system met the purchase order and specification requirements because it could run common X-devices made by the rest of the industry. If the customer had informed him of all these special requirements when the order was placed, they would have been considered in the system's design. However, such a system would have been quoted at a much higher price. He reminded the project manager that when they had specifically asked for clarification of apparent contradictions in the supplied data about the double-filleted back bend, through the sales manager, they had not received any information which warned them of this impending problem. They even had a memo for record which the sales manager had written to this effect.

There was plenty of blame to go around. The meeting adjourned at seven in the evening, not because any action plan was decided upon, but because of hunger and exhaustion. Candy bars, vending machine coffee, and threats of lawsuits had replaced the paté, vintage wine, and fellow feeling with which the project had begun.

They did not, of course, sue each other, because everybody had too much to lose if they did. The vendor knew that litigation with one of the largest companies in the industry he served would gravely damage his competitive position. He did not, however, intend ever to do business with the customer again and would not spend one penny more than absolutely necessary to get the kludge out the door.

The project manager was in no position to admit just how great a calamity he had overseen and was therefore not about to tell top management that they should neither pay for the system nor allow it to be shipped. Top management did not push too hard because they could not admit that they had spent many thousands of dollars and two years, and had received nothing of value in return.

It was in everyone's best interests to blame the technology and not their own management errors. The solution which evolved over the next few months did just that. It was announced that because of the extremely advanced technology involved, rather than attempt to implement the system in production, it would instead be delivered to a special laboratory (formerly a storeroom) where it would be perfected and optimized before being introduced into production. The vendor agreed to accept his losses, and the customer agreed not to mention the performance guarantee ever again. By conspiring to ignore the failure of the project, no one had to deal with the consequences of the failure. Therefore, everyone was satisfied, although both companies were gravely injured.

PERSONNEL SELECTION AND DEVELOPMENT

Nothing gets done unless someone does it. Plans, budgets, schedules, reports, and money—none of these will make a project a success. They are the tools which competent people, properly motivated, use to produce success. Like any tools, in the hands of incompetents they will lead only to mistakes, misdirection, excuses, recriminations, and failure. The selection of the best people and, equally important, the correction of mistaken judgments about people, are basic elements in the success of any enterprise. The technical plan provides the basic rules by which the project will be conducted and therefore provides the essential standard against which personnel performance is measured. However, by the time the feasibility study has been conducted and the plans and proposals have been completed, the professional competence of those involved should be obvious to any competent manager who has himself the necessary skills and knowledge to conduct the project. No one can manage something he does not, cannot, or will not understand, because without understanding of the questions involved, one cannot tell the competent from the merely plausible.

There are other reasons why the project manager must start with a solid understanding of the technical basis and background of the project. He cannot control a program based only on the information and insight communicated to him by those doing the work. The competent will not respect him and the incompetent will mislead him. He must see the work being done, not just glance at it on infrequent and well-planned visits to the work site, but actually watch it being done and participate in the doing at critical stages.

This does not mean that he does not delegate both responsibility and authority to those doing the work. But there is a difference between delegation and desertion. The acquisition of insight cannot be delegated, unless the right to make all of the decisions dependent upon that insight is also delegated. The subordinate who does the work which gives him the insight needed to make a decision should make it. He should have to explain and defend it to the manager, but he should make it. The manager is not a one-man band trying to do everything himself and keeping subordinates from contributing and developing. He must frequently and informally get out into the lab, look, ask questions, *listen* to the answers, make suggestions, and *listen* to the responses to them.

The opposite extreme from this sort of management is to arrange inspection visits the only purpose of which is to inflate the manager's ego. Such visits require that all work be stopped several days ahead of time so that the lab can be cleaned up (painted is even better). During the visit the manager should ask preplanned, inappropriate questions whose only purpose is to impress and intimidate the troops. Because the answers are irrevelant to the purpose of the questions, he should be sure to ignore the answers, or better yet, stride purposefully away in the middle of the answer announcing that his busy schedule does not permit him to waste any more time on such a trivial project.

There is no substitute for management insight. One does not have to get personally involved with those doing the work if such is not one's management style. But personal involvement with what they are doing, with the day-to-day work of the project, is essential. Without insight the manager will be unable to know if the project is failing until it is too late to save it or minimize the cost of failure, and he will probably be unable to find out why it failed.

Failure to meet goals will always be attributed to unforeseeable and unavoidable technical problems by those who are responsible. Sometimes this is true, but sometimes it is an excuse for incompetence, stupidity, and laziness. Those ultimately responsible for the success of the project must have sufficient discernment about it to be able to recognize the true reasons for failure to meet goals. Allowing a plausible incompetent to continue to complicate the technically simple, or firing a competent engineer who has found, faced, and tried to solve a difficult problem, can do great damage to the project and the company. Without insight, a manager cannot tell the difference between them.

The development and implementation of improved manufacturing technology is not a single expensive and disruptive step which a company completes and then harvests. It is a continuing process by which, like a body sequentially replacing each of its cells, the company continually renews its technology and preserves its competitive position.

There are two ways in which a company can improve its manufacturing technology capabilities: steady, incremental improvements to existing processes and operations, and the development of new processes and operations based on more advanced technology. The line between them is not sharp. New operations can be introduced into existing manufacturing processes so that eventually the entire old process is replaced. The incremental improvement of the new advanced technology process begins almost as soon as it is implemented in production. The company that does not have the in-house skills to steadily improve its existing technology certainly does not have the skills to develop and implement the new. The converse is also true.

Of course, once a company has a successful manufacturing technology department, such as the one described in Chapter 2, in operation, the overall staffing problem which the company in the case study faced, will not occur. But if no such team exists, or if the existing department, when evaluated upon implemented results, is not successful, top management has a very difficult problem which it must face and solve before any money is committed to any project.

The conduct of the development program is central to continuing the process of manufacturing technology renewal. Development projects are seen as large, expensive undertakings, stretching over months or years. But the small, quick job, aimed at improving a specific operation by eliminating a bottleneck or quality problem, is equally important. The line between these small improvement projects and troubleshooting is also blurred. The true picture of manufacturing technology improvement is therefore a continuum from the smallest process improvement to the largest high-tech development project.

All are equally important to gaining a competitive advantage and maintaining it. In the long run, a company must have the resources within itself to control and improve its manufacturing capability if it is to survive. Advanced manufacturing technology programs must develop the specific technology at which they are aimed, but they must also develop the technical and managerial skill base needed to perpetuate the improvement process. For this reason the hiring of outside consultants, for other than the solution of specialized and clearly defined technical problems, is a mistake. When such a consultant is retained, he should work closely with the company employees most concerned with the problem and not at cross-purposes or in competition with them. In this way, the company's skill base is enhanced, a much more important result than the solution of any one problem.

If the company, through previous long-term mismanagement, does not have the necessary resources to improve its manufacturing capability, and consultants must be hired, their assignment must be to train the company's people to do the job on a continuing basis, not to do it themselves, temporarily, and then leave.

This manufacturing technology skill base is critical to the long-term survival of the enterprise and is one of the most important assets any company possesses. All levels of management must treat it as such. Tragically, it is sometimes allowed to decay because top management recognizes neither its existence nor its importance. This is often the case in companies afflicted with wandering minstrel upper management, which comes from other companies and often other industries, stays a few years, and then leaves to undermine the long-term competitive position of some other company. The skill base takes years to build, but unfortunately it can be destroyed quite quickly by policies and decisions which do not take its importance into account.

The case study illustrates a typical personnel dilemma often faced by an upper management wishing to conduct its first formal advanced manufacturing technology project, and desperately trying to do it correctly. If asked, the company's management would have pointed to the project as proof of their commitment to advanced manufacturing technology development. This resulted from the mistaken belief that making a commitment to reaching a goal and committing funds to reach that goal are the same thing. They aren't. Commitment to reaching a goal means providing all necessary resources, not just money, and facing and solving all of the problems encountered in reaching that goal. The problem which was never faced and solved was the basic one of staffing the project.

Assume that all of the concerns about personnel which led to the program being staffed as it was were true. The manufacturing engineer was too valuable working on the old technology to be allowed to work on the new, and he did not have all of the managerial and technical credentials required for such a major project. This did not mean that he should have been excluded from the project entirely. His insight was desperately needed and his reaction to being excluded was predictable. Compared to keeping an old technology production line running, with all of the inherited technical constraints and day-to-day schedule hysteria involved, working on a new system from a clean piece of paper is an engineer's dream. By excluding him, management won his ill will for the project and all of those who were not excluded. His reaction was unfortunate, but it was understandable. He, more than anyone involved in the project, had earned the right to be included by years of dedicated hard work. Of course, he resented the project manager. This resentment could have been prevented, even if a manager were brought in from the outside, by explaining to the engineer what was planned and why, and telling him that he would have a key role in the project as the technical associate of the new project manager.

When the project manager was hired, he should have been told that he was expected to cooperate with the engineer. He should then have been assigned to work with the engineer, under his supervision, in the production shop for a substantial period of time, perhaps several months. It should have been made clear to both of them that the technical responsibility for the old technology then running, and for the new technology when it was running on the production floor, would be the manufacturing engineer's. The responsibility for planning and conducting the feasibility study and the development of the new process would be the project manager's, but both would be deeply involved in both phases. Neither could do his job without the close cooperation and support of the other, and each was expected to cooperate with and support the other. This should have been explained as a firm management requirement both were expected to meet.

When the project manager made his visits to possible vendors to assess their capabilities, the manufacturing engineer should have accompanied him. His information was essential for the vendors to prepare technically realistic proposals. He also would have provided a valuable second opinion of their capabilities, rooted in the realities of the production shop.

Remember, too, that if no vendor could be found who was truly capable of developing and building the system, the project manager had no job. He was not exactly the sort of unbiased investigator to trust alone on this mission.

When they had completed their evaluation, the project manager should have been required to prepare a proposal to management for the next step, which probably should have been the conduct of parallel feasibility studies. The top manager should then have met with them and listened to the presentation of their recommendations in detail. He should have been as interested in the way the project manager and the engineer responded to each other and cooperated in answering his questions as he was in the content of the presentation. By now they would have worked together for several months and been on the road together, traveling, eating, and working closely together. If they were becoming a team, it would be fairly obvious, as it would be if they were becoming enemies. No matter what results they presented, the project should not have proceeded unless a working relationship based on mutual respect for each other's competence and integrity had been established. It was top management's most important function to assure that this relationship had been established before committing any significant resources to the project. There is no way management can determine if this critical team building is happening from reading a management summary. Top managers, too, must personally acquire insight about the part of the project about which they must make decisions.

In the preceding parable, I have referred to the person with the most understanding of the company's present manufacturing facility as the manufacturing engineer. He could have been a first- or second-level production or maintenance supervisor, or he could have been from a process control laboratory. Obviously, every company has someone with the required insight into the manufacture of their products by the existing technology. If they did not, they would not still be in business. Top management must know who these employees are and assure that they do not become so indispensable to the daily operation of the business that they cannot be involved in advanced manufacturing technology projects.

There are several valuable side effects of this project involvement. First, because these key people will no longer be instantly available for production emergencies, it forces the organization to develop additional people with the requisite skills. This can be very important to the company's future if the key person is hit by a truck or stolen by a competitor. The second useful by-product is that it helps to keep these people from being stolen by competitors, because involving them in man-tech projects acknowledges management's appreciation of their competence and, if done correctly, will be seen as a reward.

One of the major problems of advanced manufacturing technology projects is the divisiveness they often cause within companies. This is almost always the result of management's failure to confront and solve the problems of staffing these projects. Suppose that top management does give the project manager and the manufacturing engineer an opportunity to work together and then evaluates the relationship that has developed, and it is obvious that they cannot stand each other. Suppose the project manager sees the manufacturing engineer as a grubby-handed, closed-minded mechanic who is only interested in preserving a reputation based on 30-year-old technology by preventing any technical change. Suppose the manufacturing engineer sees the project manager as a paper-shuffling Mandarin with no long-term commitment to the company and no talent, skill, or interest in the technology upon which the project is based, except as a source for buzz words with which to impress top management.

How does top management deal with such a situation? First, by not ignoring it. Personal conflict between those who are supposed to develop and implement advanced manufacturing technology has resulted in far more project failures than have unsolvable technical problems. Someone in the organization with the authority to deal with these problems must do what is needed to discover them, and then he must solve them. These are among the least pleasant and most difficult problems a manager must face. Blaming all such problems on "failure to communicate" and telling the antagonists to try to get along in the future is usually an evasion of the job the boss is being paid to do.

In the situation just described, there are four possibilities, and each requires different action by management. In the first, both the project manager's view of the engineer and the engineer's view of the project manager are false, and they are both competent professionals. The "this is just a failure to communicate, now play nice and don't fight, children" technique may be appropriate if the animosity caused by their unfortunate perceptions of each other can be overcome. But it will take more than a passing admonition from management to assure that they do, eventually, work well together.

There are three other possibilities, and this technique, which may work in the first, will not lead to cooperation in any of the others. Two of these possibilities are that one or the other of the antagonists' view of the situation is correct; one of them is a competent professional, the other is not. One must be very careful to avoid the assumption that the most pleasant person is the most dedicated and competent. The converse is often true, but there is a human tendency to punish unpleasantness much more than incompetence, and to punish disloyalty to the boss much more than disloyalty to the best interests of the company. However, it is very difficult to hide incompetence from the competent for very long, assuming, of course, that they are not willfully ignoring it because they do not have the nerve to deal with it. The most common cause of this personal cognitive dissonance is when the manager who hired or promoted the problem employee must admit that he made a mistake and replace him for the good of the project. This situation must be faced and dealt with, though. If it isn't, the competent employee will usually find another job rather than tolerate the conflict with an incompetent whom he will not see as merely being tolerated, but endorsed by management.

The fourth possibility is the very worst for the chances of the project's succeeding: two incompetents quarreling and forming factions against each other. While in the first three cases the project has some chance of succeeding, because a dedicated professional can compensate for an incompetent (if he doesn't chuck it and quit), in this case the project cannot succeed because, quite literally, there is no one to conduct it. It must be completely restaffed.

Another method for avoiding these personnel problems is to restructure the project and the areas of responsibility of the various people involved. If a baseball team has a shortstop and a second baseman neither of whom can field a ground ball or will back up the other, having them swap positions probably will not win many games.

Still another method of ducking the problem is to replace one or the other antagonist, not based on their relative competence, but on affability. Replacing the competent person on the project, because he is not able to contain his aggravation, with a second affable incompetent will spare management from the need to deal with a continuing problem and admit that a mistake was made in the original staffing. But it certainly will not contribute to the ultimate success of the project, or the company.

All of this is, of course, a simplification of what can be a complex and sometimes subtle problem. But often there are not nearly as many shades of gray as are pretended to exist in order to avoid having to deal with this basic problem of poor personnel selection.

The project team must have all of the requisite skills, both technical and administrative. Once such a team is in place and has successfully completed a project, then it certainly should be used again and again. This is the origin of the pinball game rule in man-tech development. If you want to play the next time, then you have to win this time.

The other common failure in project team building is the failure of the customer's project staff to work, on a day-to-day basis, with the vendor's people in the development of the system. This is very common in projects such as the case study, where the project manager and his staff see their duties as essentially administrative and feel that the technical conduct of the project is the vendor's exclusive responsibility. The vendor usually cannot do this alone even if he does have all of the needed technical skills available to him. If the customer's people had been intimately involved in the project, the motor drive failures would have been promptly admitted and solved. As the project proceeds, many seemingly minor problems and questions about the product the system will make and the factory it must operate in will arise. They must be answered by someone who has precisely the correct answer. They must not be guessed at or approximated, because wherever this misinformation or partially misconstrued data is incorporated into the design of the system, it will complicate its implementation and reduce its value to the company.

In the case study, the double-filleted, back bent X-device was used to illustrate this. This was a gross failure to convey the necessary insight to those who were building the system. Much subtler misunderstandings are possible, but their effects can be just as disastrous. The only way such misunderstandings can be prevented is for someone from the customer company to be frequently on the scene at the vendor's facility, participating in, not just observing, the project. If this is the right person, the vendor will know that he can get prompt, accurate answers from him or through him and therefore will not hesitate to ask. People often guess rather than ask if guessing seems easier or asking has not produced any results in the past.

No one but the project manager was involved with the project, except incidentally, and he was only a spectator. The functions he performed were surveying potential sources and negotiating, reviewing, and attending demonstrations. Advanced manufacturing technology development is not a spectator sport, and the company had no one actually playing.

The technique I have suggested here of having the manufacturing engineer continue his production support duties, yet be involved in the project, and having the project manager work on the production floor enough to gain some insight, may seem to be a violation of the management principle of assigning people specific duties and hierarchical positions. It is. Making a single person responsible for the accomplishment of a goal is not the same as assigning a single person to each task. Responsibility for a given goal cannot be assigned to two people, or no one is responsible; but many people can pitch in to do the actual work.

This primitive compulsion to divide everything into neat little boxes—work, goal responsibility, and hierarchical position—just will not work in advanced manufacturing technology development, and anyone who cannot understand the abstractions that result is probably not going to make it in this business. The worst person to have involved in these projects is someone who insists on every type of work being assigned to a specific specialist and who has a mental list of those jobs which are beneath and above everyone involved. About 80 percent of the things that are critical to project success will be found to be just outside of someone's little box.

One person is assigned responsibility for accomplishing each goal, but whoever is available and capable does whatever must be done to reach it. However, he who has goal responsibility makes sure that someone does everything required to meet that goal. Everything.

THE PRELIMINARY MEETING

A preliminary experiment and feasibility study have been completed. A program plan has been prepared and a proposal has been submitted and approved. The first step in actually doing the job is to get everyone involved, both company personnel and subcontractors, into one room and make absolutely sure that everyone understands what is going to be done, why it is to be done, what his specific tasks are, what resources he has been provided, and when he has to have it done.

Do not just send everyone a copy of the final project plan and expect them to all work separately until the various parts come magically together like a movie of an explosion run backwards. It is extremely important that the project manager conduct this meeting personally and that everyone involved be present, so that he can discover conflicts and move early to settle them. It must be run according to a preannounced agenda so that everyone comes prepared and so that weaknesses in the project team can be detected early. Those who are weaknesses will be prepared only to the extent of having excuses for not being prepared.

At this meeting, the operating budget and schedule should be reviewed and the points at which future reviews of the various tasks will be held should be agreed upon. The reviews will take place before scheduled task completion. This assures that everyone knows that he is not to keep plugging hopelessly until he is out of money and time if he finds a significant obstacle to meeting his objectives. The project manager must make it clear that he is to be informed of any obstacles as soon as they are found, so that they can be dealt with rationally. Do not ever tell someone that you only want to hear that he has met his objectives and that you do not care how he does it. This is to invite dishonesty and the chaos that always results. Accomplishing objectives is all-important, and no manager should accept excuses based only on effort, such as "we worked overtime on it and it's still impossible." (If it is impossible, why did you work on it at all?) But no one can understand or evaluate results unless he understands how they were reached.

This "I don't want to hear about your problems" technique of management is supposed to project an image of forceful leadership, but it is usually a symptom of the worst form of mental laziness and indifference to the success or failure of the endeavor. How can one have any real interest in the project and not be intensely curious about the problems it is being conducted to solve? When a project succeeds under this sort of manager, it is in spite of, not because of, him.

All questions about the project should be thoroughly discussed. At the preliminary meeting everyone should agree with the way the project is planned and accept his responsibilities and the resources provided to meet them. Then everyone should commit himself to doing the job. There are those who, although they can find no further basis for objecting, will always try to leave themselves some excuse to keep in reserve in case it is needed. Do not let them get away with it. Either they sign up or they sign out. If anyone has any rational objections to the plan, or his tasks, or the acceptance criteria for them, or the resources assigned, listen to him and settle the problem. But if, after everything is settled, someone is still trying to provide himself with an escape hatch, get rid of him one way or another. It is not fair to those who make a commitment to the project to allow anyone else to do less. It is far better to get rid of someone at the beginning of a successful project than at the end of a failure.

Not holding a detailed preliminary meeting with the vendor was an important mistake of the project manager in the case study. It was at this face-to-face meeting that the chief engineer's questions about the exact configuration and material of the production parts would have been answered, rather than evaded, as they were by being transmitted verbally by the sales manager. Transfer of technical information through nontechnical intermediaries is impossible. No important technical information is ever transferred simply by one question and one answer. The original question is usually answered by another question, and this process continues until both parties have agreed on what the question actually is. Then an answer is formulated and communicated by a similar iterative process. Attempting to do this by having a purchasing agent ask a salesman who asks someone who might know an answer is a waste of time. Any information which filters back through such a route will be useless, even if miraculously correct, because no one with any brains will trust it.

It is only with everyone sitting together that this process of transferring information, and assuring that it was correctly transferred, can take place. It must be done if the project is to succeed, and it is hard work, demanding preparation and concentration.

SUBCONTRACTOR RELATIONSHIPS

Successful advanced manufacturing technology programs are almost always conducted by teams of professionals who become intensely involved in the project. According to a report funded by the National Science Foundation[3] the development of team spirit based on trust and respect between both vendor and user personnel may be the most critical element in the ultimate success of an advanced manufacturing technology project.

Whether the lasting friendships which often come from successful projects and the lasting enmity which is a major by-product of failed projects are causes or results of success and failure is not easy to determine. But the beginning of rancor in the project team is an indication of trouble which no manager should ever ignore. He should immediately explore its causes and eliminate them. This assumes, of course, that the manager is not the cause.

In order to understand this important factor in successful customer-vendor projects, we must understand what factors contribute to cooperation. The first factor is effective project management. The project manager should be a technical person from the customer company with the necessary status in his company to effectively manage the project. He should not be a junior employee with no real authority who is only a liaison (translation: messenger boy) between the vendor and an absentee management who make decisions without the project insight needed to make them. When the implementation of these baseless decisions proves impossible, the vendor is blamed and is inadequately defended by a manager too junior to effectively argue with his own management.

The manager must be a person with a technical background holding a position of technical responsibility. This qualification is needed to assure that attaining technical goals is not subordinated to reaching schedule and cost milestones. This misemphasis may occur if the manager sees himself primarily as a project expeditor because his normal duties within the organization are concerned with getting the required check mark in the required square at the required time. In such a case, the vendor may meet schedule and price goals by evading important technical issues.

This effective manager must have technical competence, organizing skills, and personal integrity that are respected by all members of the project team. This respect is best gained by managing the preparation of a project plan based on technical and economic reality, and then assuring that subcontractors are selected by a process that is both intellectually and ethically honest. The best public relations device is to really understand what must be done and demonstrate both the skill and the character to get it done.

The second important factor in the building of successful vendor-customer project teams is equality in both skills and effort among the members of the team. Both sides must assign people to the project who are willing and able and have the time to contribute to the project. No one's duties should end with standing around with a clipboard. The vendor must supply people who have the necessary insight into his technology, and the customer must do the same. The people the customer assigns must know as much about his products, their manufacture, and their requirements as he expects the vendor's people to know about the vendor's technology. An unsupported junior engineer, who has never spent the time in the customer's shop necessary for him to understand the requirements of the products it will make and the factory it will operate in, no matter how well educated or willing, can never be an equal member of a successful project team.

This concept of equality has another facet; both sides should learn as much from each other and from the project as possible. This means there can be no we-they secrecy. If there are proprietary information problems, get the lawyers involved or do whatever else is needed to reach an agreement at the start, sign it, and then communicate. If an agreement allowing completely free exchange of project information cannot be reached, reconsider the premises upon which the project is based. If a given vendor and customer cannot reach an agreement to honestly and fully cooperate and communicate, they probably cannot do anything together very well.

The problem of vendors' proprietary information may be particularly difficult in projects where a single customer is involved with two or more vendors in parallel competitive feasibility studies or development programs. The equipment suppliers have an often legitimate fear in such situations that the customer's project personnel will, either willingly or inadvertently, transfer critical information from one subcontractor to another. The question of inadvertent transfer should be resolved by assigning a separate principal investigator to each subcontractor. Principal investigator A should never visit subcontractor B's facility and vice-versa. This should be explained to the subcontractors at the start of the program, explaining that it is being done in anticipation of their honest concerns.

Another honest concern about parallel competitive subprojects that must also be addressed at the beginning is the problem of intentional transfer of technology between subscontractors. Before it is addressed with the subcontractors, it must be addressed within the project manager's own head, and then with his management.

There is a certain rapacious beauty about letting two or more vendors work on the same problem using the different approaches that their varying technologies dictate, with the understanding that the one who finds the best solution will win a large contract. Then, one takes the best of each vendor's work and either builds the system oneself or hires a subcontractor to do it. This vendor will not necessarily be the one who had the most success in the competition, but will be the one who offers the best price and delivery for the system. Putting morality aside, and it should not be, this is one of those alluring, sharp business practices which have great attraction to people and organizations that have no long-term future, because it will assure that access to outside technology is cut off for years to come.

The manager of a project which will involve several vendors in a competitive effort should explain to each of them the rules of the competition. He should also make sure his own management understands the rules, because if they later decided to violate them, it will be the manager's word that will be broken and his reputation that will be tarnished. This is as it should be. As a professional he should understand that if he makes a commitment, he cannot then violate it and shrug it off by saying that he is acting under orders. If he makes a deal and then fails to live up to it, it is his fault and his professional reputation deserves to be tarnished, as it surely will be. Agreements should not be made without assuring beforehand that management will allow them to be fulfilled.

The equipment vendor also has a responsibility for building an atmosphere of mutual trust. The most common failure in this responsibility is not providing honest estimates of the technical capabilities of the equipment or of how long it will take and how much it will cost to implement those capabilities. Sometimes these overly optimistic estimates are the result of organizational self-delusion. A company really wants to work miracles and convinces itself and its customers, who also desperately want to work miracles, that they can be worked. This very often happens when no preliminary feasibility study provides the connection to reality needed to write and evaluate proposals.

But even when a feasibility study has been done, and particularly when the results have been very encouraging, it is important to keep a firm grasp on reality. It is far better for a vendor to promise the realistic minimum and deliver much more than to promise too much and deliver less.

This pressure to promise too much is intensified if the potential customer gives the impression that his primary concern is to get the very lowest price. If, however, the customer clearly states in his request for quotation that the vendor's explanation of how he plans to do things and how they will work is as important as how much it will cost, a much more honest proposal will result. Sometimes subcontractors promise more than they can deliver at the request of those running the project, because they must inflate the project's promised payback and reduce its difficulties in order to obtain management's approval to begin.

When, during the course of the project, it becomes obvious that the promises cannot be met, the project manager then blames the vendor and claims that he was misled. He, of course, does not admit, possibly even to himself, that he wanted to be misled and that the vendor had been somehow informed that if he wanted the contract he had better mislead him. His management, of course, helps him to attack the hapless vendor, because the alternative is to admit that they, too, were easily misled and in addition have a major problem with the project manager.

This little game of "Gotcha" is usually played more subtly than explained here, but the basic rules are the same. It is a no-lose situation for the project manager and a no-win situation for the subcontractor; if the latter does not play, he will not get the development contract, and if he does, he will be blamed for the project's failure.

The only way to avoid these disasters is for everybody to honestly prepare proposals and to honestly evaluate them. If, in the preparation of a proposal, or in the communication of any other information, someone implies that he wants a little bit better than the truth, do not give it to him. Tell him the truth. If he cannot deal with it, that is his problem, and it is best for all concerned if he chokes on it right then. Do not lie to yourself or others, for, to paraphrase Talleyrand, it is worse than a sin; it is a mistake.

Sometimes cooperation based on mutual trust between customer and subcontractors is intentionally discouraged because of the concept, which grows out of the American legal system, that only an adversary relationship can assure justice. This has been perverted into the concept that only an adversary relationship can prevent corrupt practices. Putting aside the fact that the American judicial system does not work all that well, this concept has come to mean that there are only two possible relationships between customer personnel and subcontractors: antagonism or corruption.

There is, of course, an alternative: honest cooperation between professionals representing the best interests of their companies. This is a far better program management concept than assuming that two crooked incompetents will produce benefits for their companies as long as they battle each other and do not cooperate.

If only an adversary relationship will keep two people from conspiring against their own companies, the solution is not keep them from cooperating, it is to replace them with people who can work together for the mutual good of the companies they represent. Because many people will behave up to or down to the level they perceive is expected, there is no substitute for leadership by puritanical example.

The existence of friendship and mutual respect does not mean that professionals cannot drive very hard bargains with each other for the benefit of their companies. In the case study, this hard bargaining and detailed definition of technical questions, criteria for task acceptance, and agreement on the rules by which the project would be conducted never took place. In most failed projects it never takes place. When a question with uncomfortable ramifications occurs in a failing project, it is usually ignored. It must be addressed immediately, and a reasonable plan of action to deal with it must be decided. Cooperation, if it is to benefit the project, should not be a conspiracy of mutual delusion.

PROJECT CONTROL

To manage something is to control it, and in order to manage an advanced manufacturing technology development project, one must understand and apply appropriate controls to the various elements which comprise the project. Each of these elements must be controlled with a degree of rigidity to planned requirements and flexibility based on feedback appropriate to that element.

Control by plan is the process of comparing what is actually happening, as it occurs, to what the plan predicted should happen. Sometimes a plan is prepared in great detail, because it is required by the funding source as part of the proposal, and then is never looked at again. One of the projects from which the case study was drawn was conducted in this way. The plan was followed only at the very beginning of the project, and when technical difficulties were encountered, it was abandoned.

The elements of a project can be divided into administrative and technical categories. The administrative factors, cost and schedule, and their relationship to the technical factors, must be rationally controlled with proper degrees both of rigidity and flexibility. To simply force the project to stay on schedule and budget by ignoring technical difficulties or modifying technical objectives is no better management than allowing the project to wander aimlessly while money is spent, time is lost, and no real technical progress is made. Unfortunately, it is more difficult to recognize a poorly managed project that appears to be operating on schedule and budget than the other case.

The necessary link between schedule and cost elements of the project and the technical effort is the definition of the tasks and the criteria for successfully completing the tasks in the plan (Chapter 5). These technical requirements must never be modified in order to remain on schedule or budget. Project technical requirements should never be changed for administrative reasons without the entire project being replanned and re-justified. If the technical objectives are casually or subtly changed, or portions of them are ignored because there is no time or money to accomplish them, the project that is being conducted is no longer the one that was planned and approved. Time and cost factors are extremely important in the control of projects, but only as they relate to the technical effort. It is often much less organizationally uncomfortable to evade the requirements of the project, while remaining on time and budget, than to fall behind in these closely watched and easy-to-understand factors while making real technical progress.

Budget and schedule tracking of a task should be a continuing process, not an audit on the final day to see if the goal was met. The progress toward the goal should be monitored by the project manager, frequently but informally. However, there should be some preplanned review, usually when about 80 percent of the time or money are spent on a task, when progress is reviewed on a more formal basis. If significant progress is being made and some reasonable overrun is expected, it was probably a matter of an overly optimistic original estimate, and the additional resources should be provided from the project manager's contingency fund.

If it is found, however, that significant technical problems, which were not expected by the plan, are being encountered, and therefore schedule and cost overruns are to be expected, serious effort must be made to understand the obstacle and to develop a plan for overcoming it. A rational decision must then be made either to commit the required resources or to abandon the project. This task is a difficult one for the manager because he must meet with all those involved, keep the meeting from degenerating into a blaming contest, and deal with the essential problem. This is much easier if it is done as soon as the problem is discovered, before unsuccessful attempts to solve it have polarized people. It is much better to set up an action plan and get everyone working on a problem too early, and then find it was simpler to solve than it appeared, than to treat an emerging obstacle too casually. Often, when a problem is first encountered, the search is not so much for a technically simple solution but for a solution which will not increase cost or disrupt the schedule. That is the approach which usually costs the most and takes the longest.

The technical sections of a development project plan are estimates based on the best available information, just as are the cost and schedule portions of the plan. As additional information is acquired during the conduct of the project, the technical estimates must also be adjusted as required for the project to reach its final objective. The concept introduced in Chapter

3 of applying rigid controls to the project factors where it is appropriate and flexibility in those factors where it is appropriate, must be followed.

No modification of the final objective of a project should ever be allowed. In the project described, the original objective of a production X-device bender was quietly changed to procuring a laboratory facility. Changing the final objective of a project is a common technique for covering up failure. It must not be tolerated. For example, suppose the final objective of phase I of a project is a pilot system demonstrating the technology necessary to build a production facility in the next phase. If that goal has not been met, and the pilot plant does not meet the stated requirements, the requirements must not be changed in order to go on to the next phase on time. The problems the pilot plant was built to solve must be solved, because those solutions are necessary to the design of the production system. No objective should ever be changed without a complete rejustification of the project being required. If this rejustification is no longer possible, the project should be canceled. This decision requires a degree of management maturity which is too rare.

The input materials upon which the new system or process must operate must be defined. This definition must be based on the realities of the production shop. Although specimens designed for specific tests may be part of the criteria for task completion, no development project task should ever be proven only by the processing of specimens. Every task acceptance criterion should include the processing of actual parts from the actual input materials the system will process in production. Specimens are different from parts; if they weren't, they would not be specimens.

In the development of the power turbine shaft discussed in Chapter 5, the electron beam welding tools and techniques were developed on specimens exactly the same in materials and configuration as the actual parts, in the area of the weld (Figure 6.1).

Once the welding techniques were verified, a pilot plant welding fixture for the actual parts was designed and built. This fixture had to be substantially modified because it did not provide sufficient support for the

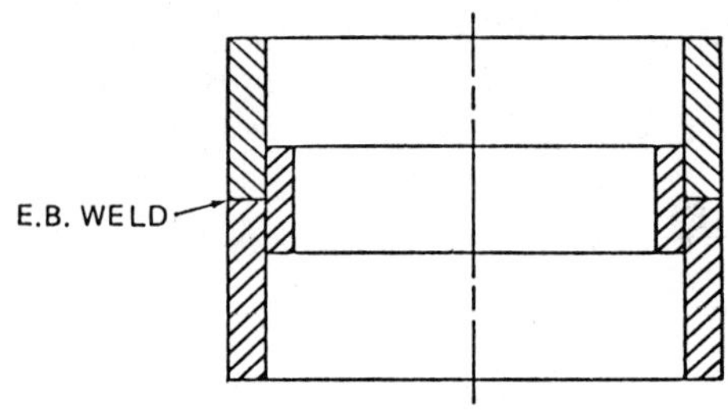

Figure 6.1.

shaft due to underestimation of its mass when assembled on the central holding arbor, and because it was ¼ inch too long to allow the electron beam to be aligned with the joint with the vacuum chamber door closed.

This project was not being conducted in the casual manner described in the case study. We were going for what the Japanese call the last grain of rice, with great care, and knew a fair amount about electron beam welding. Yet, we were misled by our experience with specimens. Fortunately, a pilot welding fixture was built early in the program, and thus the problem was found and corrected without too much embarrassment, cost, or lost time. One can readily imagine the convoluted political situation that could result if such an apparently stupid mistake were found only when the final production tooling was being tried and the door could not be closed.

Sometimes production parts from only one lot may be insufficient to detect all potential problems. During the phase II development of the laser recuperator welder described in Chapter 4, recuperator plates were shipped to the vendor's facility for equipment trials. In order to provide the thousands of parts per month needed in production, two processes, press forming and hydroforming, were both being used to make the plates. In addition, there were several sets of dies for each plate configuration. The normal procedure would have been to order some quantity of plates shipped and assume that they were representative of what would be encountered in production. The required number of plates would then have been sent, probably all from one stack made by one process and one die set on the same day.

Instead, plates were intentionally taken from every available source and lot, labeled, and shipped. They were then run to assure that they were all equally easy for the system to process. They weren't. It was found that some hydroformed plates were slightly less stiff than those from other sources. When the loading robot attempted to position these in the welding station, they would sag just enough to catch on the top of the welding tools and jam the machine. This required a slight change to the tooling which took only a few hours. Yet, if the system had been started in production, and suffered repeated jam-ups, a major implementation problem could have resulted.

The input material requirements must be developed as part of a mantech development project. These must define all essential characteristics so that, when the system is started in production, no questions will remain. The extra flange on the X-device may seem to be an exaggeration, but such things have occurred and great care must be taken to prevent them.

If it is found that the input material requirements of a new system are different from those of the manufacturing method it replaces, these variations must be carefully defined and documented. If they will have significant effect on cost, the economic effects must be immediately calculated. It is very dangerous to assume that some characteristic which is critical to the new process, but immaterial to the old, is under control just because it has always been a requirement. In production shops, specification requirements which are found to be irrelevant tend to be disregarded. Care must be taken to make any new process as tolerant of input material variations as possible, and to define and document limits based on actual tests.

As in the feasibility study (see Chapter 3), the procedure for accomplishing a task should be controlled by feedback and made as flexible as possible, while the procedures for evaluating the process and the products it produces must be rigidly specified and controlled. In poorly run projects, the opposite procedure is often followed. The planned technical approach is relentlessly pursued, no matter how inappropriate it is proved by the work of the project, and then the evaluation procedures are modified until politically expedient test results can be produced.

Development projects should be conducted to find the best manufacturing techniques, not to vindicate the brilliance of someone's original guess. Certainly, the most technically likely technique will be evaluated first, but the ability to know when to persevere, and when to admit that trial has proven an idea unsuitable and try something else, is absolutely essential to project success. This is not nearly as difficult to decide as it sometimes appears, if those conducting the project can view technical ideas objectively and not make the source of an idea the basic criterion for accepting or rejecting it. If an idea has been proven not to work, even if suggested by, or oversold to, top management, no further resources should be wasted on it. If an idea works, even if suggested by the project manager's worst enemy, it should be exploited.

The main thing which allows flexibility in the selection and development of the best technical approaches to the problems the project is being conducted to find is the building of working prototypes. If one starts out attempting to build a working production system without a prototype or pilot plant, one really has no ability to shrug and try something else. If a significant part of the resources for the full system have been spent before it is found that some other approach must be tried, chances are the project manager will not be around to see it tried. This assures that continued attempts will be made to find some way to make the original approach work, no matter how convoluted and complex. Only avowed and planned breadboards and pilot plants permit the project the flexibility to find the best solutions to its problems.

Controlled flexibility in the means toward rigid task objectives requires the precise definition of those objectives. This definition must include exactly what tests are required, how they will be run, how many will be run, and what the minimum acceptable results are. The other required characteristics of the process or system being developed must be specified with the same precision. For example, if there is a minimum production rate requirement, just stating that the system shall process a part in no more than 3 minutes, or that it shall be capable of processing so many feet of input material per hour, is inadequate. The way in which the requirement will be evaluated must also be specified. In the case of the system that is to produce a part in 3 minutes, the test should obviously not be to make a part in 2 minutes and 58 seconds and adjourn for a victory luncheon. It should be tested so that the ability of the system to meet a production schedule is proven. The test might be for the system to average 18 parts per hour for 7 hours of an 8-hour run, and that 10 such runs are required for final acceptance of the system before implementation can begin.

One of the most common causes of implementation failures is the inadequate testing of advanced manufacturing equipment under production conditions before implementation begins. The final test always comes at the end, when the project is most behind schedule and the pressure to announce a victory and ship the system is greatest. It must be resisted, and the preagreed and adequate tests must be completed with no relaxation of their conditions. If the tests uncover deficiencies in the system, which is the reason they are conducted, these deficiencies will never be so easy to correct again. A defect found in a vendor's plant, with all of his specialists, equipment, and spare parts right there, can be fixed in hours. The same defect found on the customer's shop floor may take weeks or even months to correct and severely undermine the credibility of the system. In this situation the project manager may be very alone and under pressure from both his own and the vendor's management to relax the requirements and accept the system. No matter how convinced he is of the quality, reliability, and suitability for production of the system, he should not allow any deviation from the test plan. By the time the system reaches this point, the project manager will be so involved with it that no matter how sound he considers his own engineering judgment, he should not trust it. If the system is truly ready, it is probably because the program decisions at each step were based on solid experimental evidence and not the suppositions of experts. This is not the time to stop testing and start guessing and hoping.

Suppose the system fails the final test. For example, it is to operate at 95 percent duty cycle for 15 continuous shifts producing at a stated rate and defined minimum quality level and it fails in the middle of the twelfth shift. Perhaps the quality of output falls below the acceptance level, or the system breaks down. The cause of the failure is found and corrected.

Almost always, in these situations, there will be great pressure on the project manager to run the system for the remaining three and one-half shifts, and report success.

Do not permit this to be done. The stated test must be repeated from the start and successfully completed. In the worst cases, systems barely limp through their final trials with almost continuous superficial repairs made to them, so that those involved with the project can cover up their incompetence long enough to dump its results onto a manufacturing department while dishonestly claiming that their botched project was a success.

Whether the project manager is under pressure from his vendor or his own boss to commit this outrage against his company and his professional integrity, he must resist it. The system must be proven production ready by honestly passing all required tests.

There is one other important element in the control of advanced manufacturing projects: timeliness. A company may have made a given product for years and suddenly find that its market has radically changed or even disappeared. Periodically, the project manager should find out if the effort he is making is still germane to the business realities facing the company. In some cases, such as changes in design, the project will need modification to assure that its product is what the company needs at the time of production start-up, not what it guessed it would then need, or was making, when the project was originally planned.

As with so many other things, these reviews should be well thought out and carefully conducted, and the results acted upon. This may not be easy if it is found that the spectacular new process has been made obsolete or irrelevant by changes outside of the project.

QUALITY ASPECTS OF ADVANCED MANUFACTURING TECHNOLOGY PROJECTS

Price and quality are directly related in the marketplace; better quality items demand a higher price. Price and quality are inversely related in a manufacturing shop; it costs more to make a given item poorly than it does to make it well. The assumption that the relationship between marketplace price and quality is valid in the manufacturing shop is at the root of the problems of American industry.

Poor quality performance in manufacturing increases costs. A bad part must be scrapped or reworked or used as is. The increased costs associated with scrap are obvious but in many cases are less than the actual costs of the other options.

Rework usually requires the involvement of more expensive personnel than does the simple manufacture of a part meeting quality standards. Not only must supervision and engineering people be involved, but those doing the actual rework usually require a higher skill level than is required to routinely do the job correctly. Rework, therefore, reduces productivity by keeping these skilled people from doing productive work. Rework is, by definition, nonproductive because its product is something which the company has already paid to have made. Paying for it a second time is nonproductive, no matter how the accounting department classifies the cost.

Accepting defective parts and allowing them to proceed through the manufacturing cycle is usually much more expensive than reworking them. These parts are defective, assuming that the quality standards which rejected them are reasonable. They require special care and cause problems through the rest of the manufacturing cycle, and then are released into the world to alienate customers.

All three options have another, unaccounted, but very real cost to a company; they undermine the employee's image of his own and the company's competence. In the long run, by destroying employee loyalty and fostering a what's-the-use attitude among them, this may be the most expensive result of continuing poor quality.

There are two interrelated aspects to quality as related to advanced manufacturing technology projects: the quality of the equipment developed, and the quality of the product the equipment and manufacturing process produces.

Manufacturing Process Quality

It is important to understand the concept of adequacy in the conduct of man-tech development projects. An excellent system or process, not a perfect one, should be the goal. Each attribute of the system under development should be judged on its suitability for the job it will do, and not against some contrived standard of what a perfect system would do. This striving for perfection and reaching only undependable complexity is usually the result of too many people placing too many requirements on the system. This must be resisted as much as the pressure to produce the absolutely cheapest system in the very least time. Somewhere between these two failure modes, one too shoddy to work in production and one too complex to ever reach production, is the adequate, dependable, and usually elegant system, which should be the final result of the project.

Overdesign in manufacturing systems is usually caused by trying to get the system to do more things than it should or must do. Underdesign is most often shown by components and subsystems which are not rugged enough for the service and duty cycles intended. The motor drive in the X-device bender is an example of this. It is possible and unfortunately quite common to have a system that is both overly complex and underdesigned at the same time.

Unfortunately, the management constraints under which systems are designed and built are based on a superficially rational compulsion to obtain the very most for the very least. A system which costs $100, is supposed to perform six different functions, and does not work is not cheaper than one costing $200 that does only two functions superbly.

The best systems, and the easiest to implement, are those whose design is simple and elegant and whose components and subsystems were selected and built with sufficient margin to assure that they will continue to function under far less than ideal conditions.

Often, the subsystems which cause the most unreliability are accessory functions which are peripheral or even unessential to the purpose of the system. An automobile which cannot be driven at night because no spare motor is available for its covered headlights would be an example of this foolish complication. Adequacy is neither wastefully shoddy nor wastefully complicated. It is neat.

One of the most important long-term advantages that a company can gain from a well-conducted project results from developing an inherently controllable manufacturing process. Such processes have realistic requirements for input material quality as well as operator and maintenance skills. If a process will make a part of reasonable quality only if all of the equipment is in perfect condition, the input materials are flawless, and the operator performs his tasks exactly as required, that process should not be implemented. It still needs considerable development.

A new process must be capable of accepting input material to standards that can be realistically expected. Its maintenance requirements should not be such that the least amount of wear or dirt will cause it to fail. Finally, the operator should have to expend extra effort to make a bad product, not to make a good one.

All of this, too, is a part of the adequacy of design of the new process and its equipment, and must be constantly considered during the conduct of the development project.

Product Quality Standards

The electron beam welded compressor rotor for the T-53 engine was one of the first welded aircraft turbine engine rotating components to be placed in production (Figure 6.2). Rather than being the bolted and splined assembly then common, the entire assembly is joined by three welds (Figure 6.3). At the time of its development in the 1960s, it was quite revolutionary.

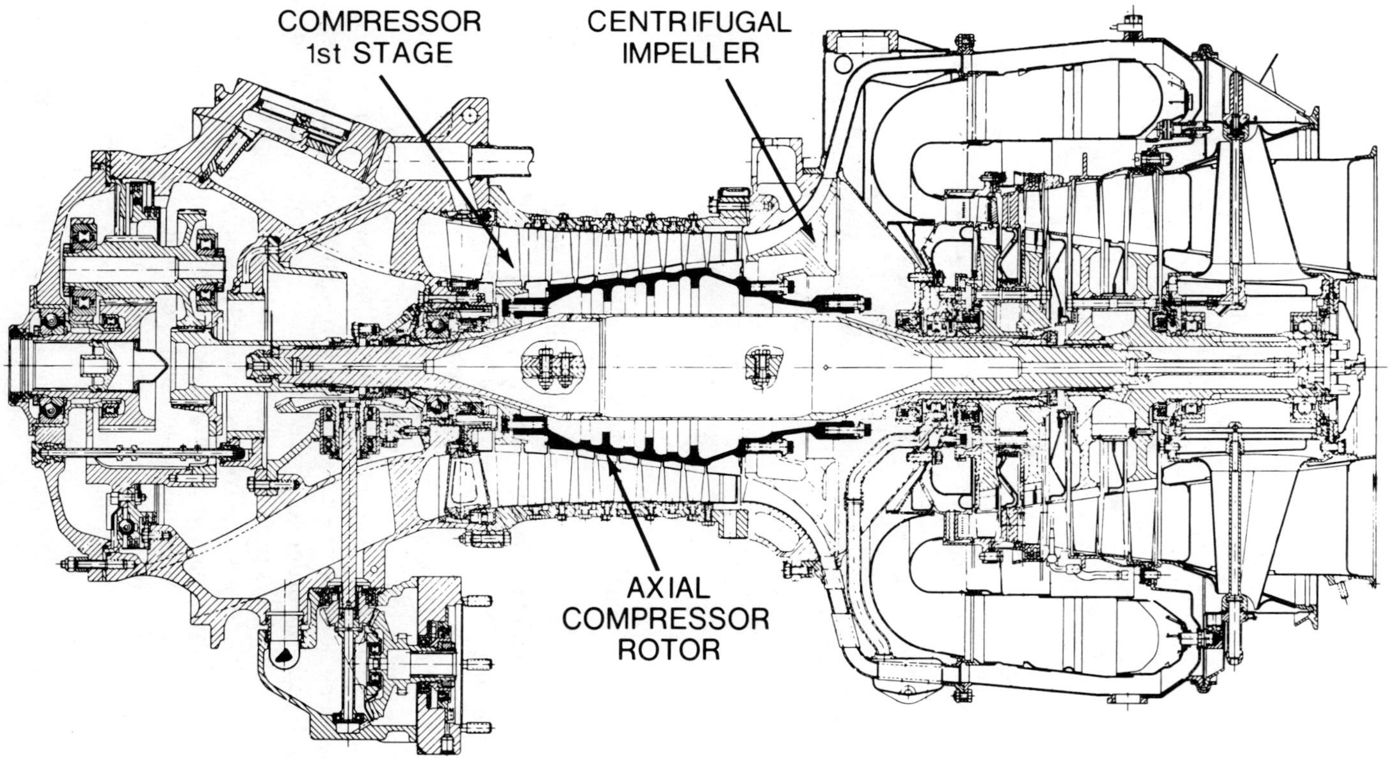

Figure 6.2. T53 Gas Turbine Engine.

Figure 6.3. T53 Compressor Assembly.

A competitor had been working on a similar assembly of the same titanium material but had been unable to produce welds of the required quality. The answer to the basic problem which caused excessive porosity to occur in these welds was found and verified by producing one perfect weld after another. The problem was not in the welding process itself, as was then assumed, but in the preweld cleaning of the details.[4]

Because the idea of trusting a weld in a rotating component of the engine for a single-engine aircraft was so radical, rotors were made and tested for several years in order to prove the basic concept. In all, about 60 rotors were made in this project and each of them had three perfect welds. These were verified by extensive destructive tests of both specimens and parts, as well as by inspecting every assembly by fluorescent penetrant techniques; by single-wall radiography, both perpendicular and at an angle to the surface; and by two different immersion ultrasonic techniques. The reason no other inspection methods were used was that there were no other inspection methods.

In order to have defective welds to develop and calibrate the inspection techniques, welding procedures to produce specific defects in specific locations were developed. When this can be done, a fair amount about the causes of defects is known.

Once the basic welding procedure was developed, every weld throughout the project was perfect. Then both the part and the process for its manufacture were implemented in two different plants, a thousand miles apart. Soon, a few welds with minute but detectable defects occurred. No one knew if these very occasional and very minor defects were rejectable or not, because all tests had been done on perfect parts. The parts with defects had to be impounded while an investigation of defect tolerance, which should have been part of the original project, was conducted.

An essential part of the development of any new manufacturing process is to establish quality control procedures and acceptance limits for the product. Obviously, this must be done in a case such as the compressor rotor, where both a new part and the technology for its manufacture are developed. However, even when the new process will make a long-established part, with known limits of acceptability, these quality aspects should be investigated for several reasons. It may be possible, using the new process, to improve the quality of the product and to enhance its competitiveness if it is a finished product. If it is a component of a product, improving it may simplify later assembly operations or eliminate or reduce inspection operations. If there is no gain from increasing the quality requirements, do not do it just because it is possible. Leave the old requirements in place, and let the added quality margin assure that there will never be a reject.

Inspection techniques for the new product of the new process must be developed as part of the development of that product. In the case of the compressor rotor it was not known what would, in the long run, prove to be the most dependable or suitable technique. There were several experts at the time who were sure that a partially unfused electron beam welded joint would be so tightly pulled together that fluorescent penetrant or radiographic techniques would not detect the resulting massive defect. Therefore, the ultrasonic techniques were developed. Attempts to make such a joint proved that it was not possible. Although the joint faces are machined to high-quality finish, they are not optical flats and there is enough inherent roughness to assure that both X-ray and fluorescent penetrant will always find this defect. Yet, the part and processes were so advanced that all of the inspection techniques were used for over 2000 assemblies before the redundant inspections were eliminated.

This undoubtedly is the ultimate in development project inspection paranoia, but it was justified in the case of a new aircraft engine part made by what was then a very advanced process. However, the approach used, making the development of inspection techniques and procedures for verifying the quality of the new component an integral part of the project, should always be followed. In the case of critical products, the successful development and verification of the inspection procedures is an objective that must be met or else the new manufacturing process, no matter how apparently successful, cannot be used in production.

Most projects will not involve a product with such high quality requirements, but the procedures for controlling the new process must still be developed and proven as part of the development procedure.

Quality control procedures based on inspection by assumption should be avoided wherever possible. In these, a sample is periodically tested by a destructive technique and, if found acceptable, it is assumed that the parts entering service are acceptable. This technique is only as good as the controls which assure the correctness of the assumption that the inspection sample and the product are identical.

It has unpleasant implications when a sample is rejected. When that happens, every part made since the last acceptable sample must be either scrapped or reworked, because now they must all be assumed to be unacceptable. Most inspection-by-assumption systems, left unaudited to operate in a production shop, will eventually collapse. Usuallly the making of the samples will be subtly modified to reduce rejections and thus destroy the validity of the assumption. This inspection concept is obsolete and should be avoided whenever possible. Techniques for controlling the process itself and for nondestructive inspection of the products actually entering service are much preferred.

REPORTS AND DOCUMENTATION

The first question to be decided about written reports is how frequently they should be required. If they are done too frequently, such as on a monthly schedule, people spend too much time describing what they are doing and too little time doing it. Another problem with such frequent reports is that they force people to concentrate on the quickest and easiest parts of the project because they must constantly generate something to report.

If, however, the reporting period is too long, such as every six months, or only at the end of each major phase, the project may wander far off course before being evaluated. For these reasons, interim written technical reports every three months during the conduct of the project, as well as a major report at the end of each major phase and, of course, at the end of the project, are probably best.

Reports have two important functions: first, to enlighten the author, and second, to enlighten the reader. The preparation of a report, if it is done correctly, forces logical ordering of the information. This will often detect contradictions or missing pieces of information where one thought one's knowledge was complete. This process of improving one's own understanding of the project by presenting it so that others can understand it is either enhanced or impeded by the format of the report.

If the approved format has page after page of required verbage, in which the actual report is immersed like a few raisins in a barrel of gruel, flaws in the logic of the project will not be detected. The format of periodic written reports should be simple and concise.

A one- or two-page background section, which is reused each quarter, should be followed by a description of what actually was done during the reporting period and the results. All of this should be explained in terms of the CTQs, tasks, and task criteria of the project plan. It should explain the significance of the results and what, if any, conclusions should be drawn from them. In some cases it is important to explain what conclusions should not be drawn from the results and why. The final section should explain what is planned for the next reporting period, and explain how it will contribute to the accomplishment of the project's goals.

The entire report should probably not be more than 10 pages long. It should include illustrations and tables, but only to clarify, not to confuse or pad. If the report is three pages long and concise and complete, it is far superior to 25 pages of rehashes and obscure calculations. The emphasis must be on information, not raw data, and on results, not busy work.

Most engineering reports are very poorly written, although far superior to the "unconceptualizing and decommunicazation" of the English language practiced by social scientists. Most of these problems in report writing can be solved by rigorously outlining the report before writing it, and by frequently reading Strunk and White.[5]

Poorly reasoned and sloppily executed work is acceptable neither in the conduct nor in the reporting of a project. Although the manufacturing process and the equipment it requires must be designed and evaluated to practical standards, based on the real world of the production shop, every facet of the project should be conducted to the highest professional standards. Reports filled with clichés, buzz words, typos, and gibberish are a good indication of projects based on superficial thinking, opinions unsupported by experimental evidence, and political cunning.

There is, of course, another, and probably more important, form of reporting than the quarterly and final reports; that is the documentation of exactly what was done, when and why, during the conduct of the project. This is so obvious, and so important, that it is amazing how often one needs to go back and find out why something which will not work now would work then, and finds the notes to be so incomplete as to be useless. Particularly when things are going well, there is a tendency to press on with the work itself and postpone writing the documentation. It is a temptation that must be resisted.

The final manufacturing process must also be recorded in detail. This is usually in the form of manufacturing process instructions. These should be drafted, and followed during the final preimplementation trials, to assure that what they describe and require is exactly what was done. Often, small changes and improvements are made during the final acceptance test runs. The paperwork must reflect the very last, very best, version of the process—the one that will be implemented.

Documentation of computer programs and transfer of the final version to some permanent memory are critical to expeditious implementation. Imagine having to debug a long and complex computer program on the production floor because the final, completed version was dumped during transfer of the equipment, and in the last rush to complete the project, it was never transferred to hard memory. Dumb, but unfortunately common.

The purchase order for the equipment for the new manufacturing process will undoubtedly have a standard section requiring that the vendor provide a stated number of copies of an operator's and maintenance manual, as well as various system drawings. Usually this requirement is forgotten until the equipment is delivered, at which time a stack of sketches, drawings, and notebooks also arrives. These are accepted as meeting the requirement of the purchase order without being read to find out if they have anything to do with the system actually purchased. Sometimes the connection is tenuous.

The vendor should be informed at the beginning of the construction of the production equipment that these documents are very important to the customer's ability to keep the system running in production and will be treated as such. A date for the delivery of these documents, in draft form, to the customer's project staff should be included in the project plan. This date should be early enough so that they can be reviewed in detail prior to final acceptance for shipment of the equipment. The vendor should be informed that if there is so much as one ladder diagram missing, the equipment will not be accepted, no matter how well it works. If this is not done, it may be months after the system is delivered before the documentation is complete, if ever.

PLANNING FOR IMPLEMENTATION

The actual planning and conduct of the advanced manufacturing technology project is probably, at most, half of the total job. Once a system is debugged, tested, and running, the easy part is over and the tough part begins: implementation in the production facility. During the development phase of the project implementation must be an overriding concern.

In many projects, which later founder, as in the case study, it is hardly considered. It is asumed that once a system or a new manufacturing process is tested and accepted, the project is completed and it will be set up on the production shop floor, plugged into the nearest electrical outlet, and just started. Anyone who expects this has never implemented a process in the real world. To provide someone with a violin is not most of the job of getting him to play it well.

In the following chapter the problems of implementing new systems and processes will be discussed in detail, but it is during the conduct of the development project that whatever can be done to reduce the potential difficulties must be done.

The system must be designed and proven to be able to operate on the actual input materials it must process. The example in the case study of building a system that could only process specimens was taken from an actual occurrence. There is an attraction to this walk-before-you-run approach of starting with something easier to make than the actual part and then going on to actual hardware. It is the way a new piece of old technology replacement equipment might be qualified for production where the relationships of the parts, the specimens, and the process are completely understood. Particularly if the input materials or detail parts to the new process are expensive, it is easy to justify postponing the moment when the system must do what it has been built to do for as long as possible. It is a mistake. Run parts early—only they matter. Planning for production implementation requires that everything possible be done to assure that the system is capable of accepting the actual input it is to accept and of dependably producing acceptable output.

Every effort must be made to assure the system runs dependably and consistently. The marginal motor drive system of the case study is quite common. Equipment which ran perfectly during its trials and was thought to be grossly overdesigned has been found to have components which were insufficient for the usage of the production shop. Any components which are even suspected of being marginal during the development project must be rationally analyzed and replaced. During the final debugging of the recuperator laser hole joint welder (Chapter 4), intermittent electrical problems were twice found to be caused by a particular wiring harness connector which was used throughout the facility. These were supposed to be an improvement over the older type. In both of the failures it was found that conducting pins within the connectors either had been loosened during assembly or had worked loose during operation. There were dozens of these connectors, each with many pins; but after the second failure, the vendor and the customer, working together, decided that all of the connectors had to be replaced, not just the ones which had so far failed. They were not suitable for production. So everything was shut down, and every connector was replaced with one of the older, more dependable type. Solve every problem that is found during the development project. There will be plenty of new ones when the system gets to production.

Production suitability requires that the final system be both rugged and maintainable. Here is another area where the customer should be intimately involved. When the system breaks down, whether it is a pilot plant or the final system, is exactly the wrong time for the customer's representatives to go to lunch. They should see exactly what is involved in finding the cause of the failure and repairing it, and try to envision the work being done, not by the vendor's people, but by their own maintenance department. How much dismantling is necessary? What special tools and skills are necessary? Are the parts being replaced the ones which are supplied in the standard spare parts kit? Or was the selection of parts for the kit based on other applications of different equipment models? This is the sort of information which is critical to keeping the system running in production. Get it.

Toward the end of the development project, when the new manufacturing process and the equipment are nearing their final form, the planning for the training of the manufacturing people who will operate it and the maintenance personnel who will service the equipment must be done. This training is central to the success of the new technology in production and it must not be treated casually.

The project staff and the equipment vendor should plan training in detail and not just pass it off with the vendor agreeing to give a three-day course to the customer's people once the system is installed. In particular, the training of equipment maintenance personnel (see Chapter 7) should be very carefully planned and done because inability to properly service and promptly repair equipment is a major cause of failure of implementation in production facilities.

In addition to deciding who will teach the course, what he will teach and to whom he will teach it must be considered. The maintenance tasks involved must be evaluated in terms of how the customer's maintenance department is structured and operated. Particularly in union shops, the fit between what must be done to service the equipment and the way tasks are divided between the trades may cause major problems when installation begins. These should be investigated in advance. One way to do this is to make lists of the most common maintenance tasks and have someone who understands the union contract list next to them which trades are required. If it is found that the servicing of a component which one person can hold in his hand requires that four people work on it simultaneously, the personnel department must be informed so that some rational solution can be negotiated with the union.

A question which will later cause problems in the evaluation of the implemented system is its production rate. Production rate should be distinguished from capability. Production rate is the amount of its product the system will make, on average, over a fairly long time, operating on a production basis. Capability is the short-term output of the system under ideal conditions. If someone has an automobile capable of going 120 miles per hour and lives 14 miles from work, no one expects him to commute daily in seven minutes flat. Real world constraints are accepted. Yet, if one makes the statement that a new system is capable of making a part in 30 seconds, the pocket calculators come out and someone announces that working three shifts per day, seven days a week, 20,160 parts (2 parts × 60 minutes × 8 hours × 3 shifts × 7 days) per week can be made. They then magnanimously agree to only scheduling 20,000 parts per week for the first month of operation to allow time for operator training.

Production capability and the expected production rate when meeting a schedule are totally different and must be represented as such when preparing for implementation. Otherwise, expectations so high that they are impossible to reach may make a very successful system be viewed as a failure.

When the real production rate is estimated, factors such as scheduled start-up and shut-down time, time for preventive maintenance and reloading with both raw material and expendables, quality control procedure time, actual operator availability, and any other factors found to be necessary to the operation of the equipment or process must be included in the calculations. If these data are to be used in comparison with existing systems, those systems should be evaluated by the same methods and on the same bases to assure the comparisons are valid.

Toward the end of the development project, when the technology being applied and the new procedures and processes are fairly well understood, it is an excellent idea to get everyone on the project team together and ask what they think will be the most troublesome factors in the production start-up. This was done at the end of the titanium compressor rotor project, and the answer was that, although all of the high-technology manufacturing and inspection procedures were as ready as they would ever be, the most common continuing problem was the lathe cutting of the electron beam welded joint faces prior to welding. This operation, because of the very tight tolerances required, continually produced parts needing remachining before they could be assembled for welding. The joint had originally been designed to provide the best fit possible in order to eliminate any potential welding problems.

Now, near the end of the project, when much more was known about the welding of the component, it might be possible to redesign the joint in order to eliminate some of the machining problems, without harming the welding operation. The exact amount by which the tolerances on the various dimensions could be widened was unknown, so a program to modify the design, weld parts with the various modifications, and test the resulting welds was done. Figure 6.4 shows the original and final joint designs. The final production design was significantly easier to machine and assemble. It could not have been made at the beginning of the project, because then there were so many unknown factors that the design had to be as conservative as possible. One must, near the end of the project, review it completely to see if there are things that can now be simplified because of the insight gained from conducting the project.

During the development project, in order to prepare for implementation of the new technology, health and safety problems must be considered. These usually fall into the following groups: those hazards which are expected and which codes and equipment manufacturers' designs are supposed to preclude, production shop hazards which result from defeating the safety features required by codes or included in the manufacturers' designs, those caused by mistakes so bizarre that no sane person could anticipate them, and finally, nonexistent but vividly imagined hazards.

The first class of safety hazards are normally not a problem if all of the requirements of the relevant codes are met. If, for some reason, some requirement cannot be met, the reasons, and any alternate methods selected for dealing with the hazard, must be fully documented. If it is not possible to write an explanation of why a certain safety feature was not included, or why a specific hazard was decided to be irrelevant, which will stand the closest of scrutinies by a hostile and hungry attorney, the safety feature had better be included or a defense against the hazard devised.

The actual equipment is usually not so much of a problem as its location within the plant and the hazards caused by proximity to other areas and equipment. Sometimes health and safety problems are ignored because those running the project did not consider these matters originally. Later they are afraid to raise them lest they be criticized. A particular example of these sometimes expensive safety problems involves the disposal of hazardous wastes from manufacturing processes. Particularly if no pilot plant was operated to determine the exact volume of such wastes, this can become one of the worst areas of cognitive dissonance in project management. It is a very unpleasant surprise to discover that, instead of a

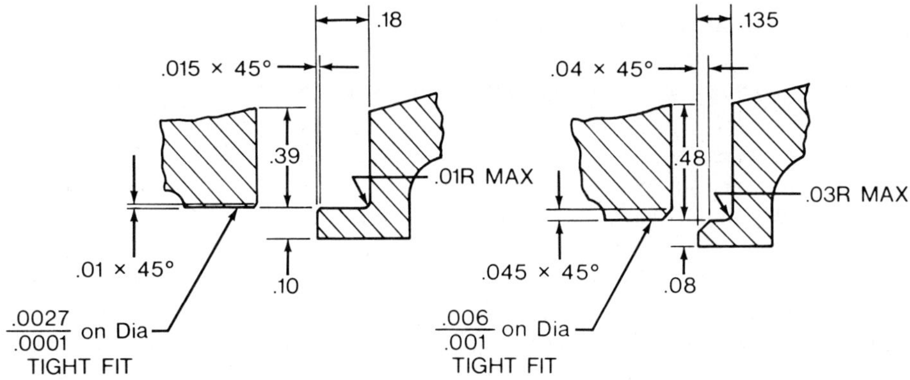

Figure 6.4. Compressor Rotor Detail Part Joint Design.

quart a month of some difficult to dispose of material, the new process is making 10 gallons a day. To deal with it as soon as it is encountered, rather than to ignore the problem by quietly dumping the material down the sink, takes considerable strength of character.

The second class of safety hazards comes about because of the terrible compulsion some people have to remove any safety device which they find inconvenient, no matter how minor the inconvenience or how major the resultant risk. People will C-clamp switches designed to assure that they cannot reach into the press and then reach into the press. About the only way this problem can be addressed is to have someone who is familiar with the characters who work in the facility where the system will be installed watch it run and try to anticipate and defeat the particular types of suicidal stupidity the machinery may meet in the shop.

The third type of safety hazard which must be dealt with is the imagined one, which nonetheless can cause very real implementation problems. I have been seriously questioned by people who had somehow (perhaps from the movie *Goldfinger*) been convinced that working in the same building with a laser could make one sterile. These fears can often become rumors which become accepted truth if ignored by project management. Find out the concerns of the employees who will ultimately make the new technology succeed or fail in the shop, and if these concerns are based on misinformation or misunderstanding, take steps to solve the problem. Do not ignore, or worse, laugh at them.

Sometimes, during the development project, those who will eventually operate and service the system in production must be introduced to it. It has been suggested that the best way to do this is to make the person in the production shop who is most negative about it work with the development people during the project in order to convert him. Converts often make the best evangelists. Even if one does not go to this extreme, those who will have the most effect on the success of the implementation should be shown the system, be trained on it, be allowed to ask unlimited questions which are honestly answered, and be allowed to make suggestions which are honestly considered and, if useful, accepted. They should be made members of the project team.

Sometimes, those who are working on the project are unwilling to involve the production people, supposedly because of their lack of background in the new technology. This may sometimes be a rationalization of an unwillingness to face questioning by those who have no romantic illusions about the project and are liable to be more concerned with how well the system actually works than with its scientific truth and beauty.

Sooner or later, shop people are going to get their eyes and hands on the product of the project. It is far better to have it be sooner. The idea that was popular a few years ago, that as the technical level of manufacturing equipment was raised the skill level of those who would operate it could be reduced, has pretty much been discredited by actual experience both in the United States and abroad. The idea was that capital equipment would not only replace labor but eliminate skilled labor almost entirely. It just does not work that way in most cases. Although less labor may be required, the skill level is invariably increased; if not in those who operate it, surely in those who service and repair it.

The need to train these people must be considered during the development project. People fear more than anything else that which they do not understand. Everything possible must be done to make them understand the new technology.

REFERENCES

1. Woolf, Henry B. (Ed.). *Webster's New Collegiate Dictionary,* Springfield, Mass.: Merriam, 1977.
2. Tuchman, Barbara W. *The March to Folly, from Troy to Vietnam,* New York: Knopf, 1984.
3. Podolsky, D. M. Vendor/user relationship plays key role in implementation. *Robotics World,* 6–8 (October 1984).
4. Miller, J. A. Manufacture of an electron beam welded turbine engine compressor rotor. *Welding Journal,* **56**(5):24–29 (May 1977).
5. Strunk, William, Jr., and White, E. B., *The Elements of Style* (2nd ed.), New York: Macmillan 1972.

7
PRODUCTION IMPLEMENTATION

INTRODUCTION

Production implementation is the transfer of the new manufacturing process and equipment from one functional unit of a company to another. Interfaces between departments are the most frequent sites of conflict in any organization, and any transfer across them is one of the most difficult of management activities. To implement a new manufacturing technology, not only the manufacturing process and the required equipment but also the responsibility for its operation must be transferred from those who developed it to the manufacturing operations department which must operate and maintain it on a daily basis.

Manufacturing supervision and management is a complex and difficult job. A middle- or first-level manufacturing manager must meet both budgets and relentless production schedules while under a number of constraints he is powerless to change. Someone else negotiates the union contract which rigidly defines his relationship to his work force. Someone else designs the product he must make and the quality requirements it must meet. Someone else decides which product to make and promises the customer when it will be delivered. Someone else sets his budgets for manpower, materials, and equipment, and these are often the first to be cut when things slow down and the last to be increased when times improve.

It is important for those who develop new manufacturing technology and are responsible for its implementation to understand the unfortunate siege mentality which daily operation under those constraints has produced in some manufacturing departments. Anything coming from any other department is assumed to be arbitrary, unrealistic, and unreasonable. For top management to expect a man-tech project manager, with no authority in such a shop, to introduce something as disruptive and threatening as an important new manufacturing process is at least naive.

The project manager must thoroughly understand the culture of the shop in which the implementation will take place. It is usually safe to assume that things are worse than anyone in top management comprehends, and take every possible step to gain the cooperation and support of shop supervision and convince them that the new technology will solve some of their worst problems, not create new ones. This is why it is so much easier to implement a technology and equipment whose development were firmly based on an understanding of the existing problems in the production shop than to implement the same basic technology developed without any concern or understanding of those particular problems. To gain acceptance for the new technology, the project manager must convince the shop people that it is not a hindrance but an advantage to them, and that those who developed it were working toward the same goals that they must meet.

The credibility gained from routinely providing technical support to manufacturing is an important asset in convincing shop supervision of the relevance of both the technology and those who developed it to their actual problems. If, in the preliminary discussions of how a new technology will be implemented, those who developed it show any lack of insight into the current processes and products, and about how the new technology relates to them, the implementation is already in trouble.

IMPLEMENTATION RESPONSIBILITIES OF UPPER MANAGEMENT

The management of advanced manufacturing technology development, as described to this point, is primarily the responsibility of the project manager. He has and should run the development project with very little day-to-day top-management involvement. The essential duties of upper management are to select the project manager, provide him with the people, environment, and resources he needs to conduct the project; and leave him alone to do the job.

But as implementation approaches, this essentially passive top-management role must become much more active if the implementation is to be done efficiently. The project manager usually only has authority on one side of the department interface which the implementation phase must bridge, and he can only succeed unsupported with great difficulty. If the project eventually succeeds because of the extraordinary efforts of the manager and the development project team, its implementation will have been far less efficient and prompt than it would have been if upper management had done certain things which are necessary and which only they can do. Only upper-level managers have authority which straddles the interdepartmental barrier the project must now cross.

They must do these things: designate someone to manage the implementation of the project; publicly grant him the authority needed to fulfill this responsibility; and publicly and unequivocally commit themselves to the success of the implementation.

These two transitions, the new process moving from the lab or vendor's facility to the production shop and the transfer of responsibility for the process from the development staff to manufacturing personnel, must never be planned to occur simultaneously, no matter how superficially attractive this might appear. Neither the new process, nor the responsibility for it, can be dumped on the doorstep of manufacturing with the expectation that they will have the time, the resources, or the interest to bring it in and nurture it as their own.

TRANSITION FROM DEVELOPMENT TO IMPLEMENTATION

The implementation phase of the project should not begin when the new manufacturing process and its equipment arrive in the production shop. This alignment of the two most fragile joints in the advanced manufacturing technology development procedure is a common cause of project failure.

The involvement of the manufacturing and maintenance departments should begin before the development project is completed, and the development project staff's involvement must continue until the production implementation is truly completed. Therefore, as the development phase nears completion, and certainly before the final acceptance trials of equipment, upper management should designate certain key first- or second-line manufacturing supervisors to join the project team. But first they must select a person to manage that team through this final project phase.

The manager of the implementation phase will be responsible for transferring the technology across the departmental interface. He must, therefore, have authority across that interface, because there cannot be responsibility without authority. This is usually very difficult for upper management to accept. The idea of interdepartmental authority is so foreign to the concept of division of tasks that many managers will not even consider it. Yet, the implementation of a new manufacturing technology is a single task and must be managed and conducted as such. Because this task happens to fall across the borderline which classically separates two departments is no reason to cut its management down the center and fatally weaken it.

The early selection, designation, and granting of authority to the person who will manage the implementation is extremely important, so that he can take all of the steps necessary to assure the prompt and efficient production start-up of the new technology before that start-up begins.

Implementations often fail because no one person is assigned to manage them. Plant engineering prepares the site and utilities. Manufacturing engineering prepares the specific process instructions. The engineering department makes the required product design changes. Perhaps someone informally coordinates these efforts without being publicly designated by upper management, and with authority only to mediate conflicts, not to decide them and enforce those decisions.

If a project of the same magnitude or importance were to fall totally within a single department of a company, someone would routinely be given both responsibility and authority for it. But because implementation straddles departmental boundaries, no such action is taken. Often these projects fail, not because they were poorly managed, but because no one managed them at all.

The normal evolution of a successful advanced manufacturing technology project results in a personnel problem that often leads to implementation failure. Projects start small; a feasibility study is a relatively minor activity. Because it is minor, and because the new technology involved is seen to be the proper, high-risk activity for younger employees, it is assigned to a junior engineer. If the feasibility study works, this person will probably play a key role in the development project, because he is now the one with the most insight into the new technology. He may be the principal investigator because he is too junior to be the project manager. But if the project is considered risky and radical, and if the designated project manager will not or cannot become deeply involved, this junior engineer will, in effect, run the project. This is considered good management practice, because it rewards initiative with a longer rope.

Failure, of course, solves all problems. But if the development project succeeds, and its successful implementation becomes important to the company, what then? Should the junior engineer be granted the interdepartmental authority necessary to the task, or should someone else, without his profound understanding of the project but with the status and maturity necessary to deal with management of the manufacturing department, be put in charge? This is a tough one.

If an outsider is brought in and made project manager at the end of a highly successful development phase, the effect on the morale of the project team, and particularly the young engineer who led it, will be devastating. In the worst case, he quits and leaves the company with a marvelous new manufacturing process and product which no one understands well enough to implement. If he stays, his performance will not be improved by having been punished for succeeding.

Yet, should a young person, perhaps in his twenties and with very little management experience, be given the authority to severely disrupt the company's manufacturing facility? It is here that upper management's understanding of the political and cultural realities of their own manufacturing shop is important. The maturity and professionalism of the young engineer must also be honestly evaluated, and this is very difficult because there are all sorts of managers who will say, "Young Blank is very brilliant and has done an excellent job on the project, but . . . ." After the "but" will come whatever they think upper management will find least pleasant. If top management tends to be supercilious, he will be described as "overambitious and pushy." If top management is overly concerned with harmony within the organization, he will be described as "not a team player." These descriptions invariably tell a great deal about those who concoct them and a top management that will believe them, but they usually tell nothing about the person in question, except that he is perceived as a threat because of his competence, his youth, and the successful results he has produced.

Competence and the successful results are all that count. He is the only person who can implement the innovation, so let him get on with the job by giving him the authority and support he needs. If he does not have the financial skills necessary, tell him so and assign someone from finance to his team to help him. If he may have trouble dealing with the older managers in manufacturing, assign one of them who is respected and about to retire to the team to help him. Discuss potential problems with him and arrange the support and assistance he needs. But do not place him in a position where he must leave the company to preserve his self-respect, particularly because he is seen as a threat by less productive but more senior managers.

Even if the development project manager does have the experience and status within his own department to conduct the implementation, in many companies he will not be granted the interdepartmental authority to do it efficiently. The reasons for this have nothing to do with him or the project but are the result of obscure pecking-order problems. This is why the concept of a project champion is so attractive. Top management does not have to face the often difficult organizational decisions that implementing advanced manufacturing technology forces upon them.

The project champion is often someone who somehow gets the job done in spite of top management's refusal to give him the recognition and authority it actually required. The usual technique is to quietly appoint a project manager but to allow him only the authority that he already possesses in the department to which he is normally assigned. He is then told to matrix manage the project and, if any problems occur, to refer them up the chain of command in his own department.

This technique has the advantage of allowing upper management to avoid delegating the authority that the project manager needs and to interfere and micromanage the project themselves. Assume the project manager is from the engineering department. All requests and conflicts must therefore travel upward through the engineering department hierarchy to the V.P. of engineering who then meets with the V.P. of manufacturing and their common boss, who discuss the problem and reach a decision. This decision is then passed down through the manufacturing hierarchy for action. The chances of such a decision being correct are close to zero. The poor project manager must then try to get this decision corrected, but he will usually settle for ignoring it so that at least the problem will not be made worse.

Of course, if the project manager is fairly junior in a company with many layers of management, his request will never reach the V.P. of engineering. It will reach some intermediate manager who attained his position by scrupulously avoiding any controversy. It will then be either pocket vetoed or bumped back down with instructions to "work it out as best you can. This is not the sort of thing I can bother the V.P. with."

This inverted U-structure of project management does not solve interface problems, it aggravates them. The task of top management in advanced manufacturing technology projects should not be to make interdepartmental boundaries more impassable but to make them disappear.

Another technique used to avoid having to grant authority to an engineer in the manufacturing shop, or a manufacturing person if he is selected to be the project manager over engineering personnel, is to give the job to a professional manager from some other department entirely. This person usually knows nothing about either engineering or manufacturing and is therefore seen as the ideal person to straddle the interface between them, probably because he will irritate both sides equally.

In the worst cases, this manager will be some corporate golden boy, selected because the so far successful project will be a nice step for him on his way to top management. His appointment will be justified by saying that the project manager who has succeeded thus far, although a talented technician, does not have the management skills necessary to run such an important project in such a critical phase. A much favored choice for the position is someone with an MBA, a financial background, and a profound ignorance of and barely concealed contempt for the technology.

The former project manager will continue to run the project, if he does not quit, because he has the technical insight and the respect of those involved needed to run the project. Meanwhile, the designated project manager will stay in his office, as far from the scene of the action as possible, issuing nonsensical memos and demanding detailed justifications for the most trivial and obvious expenditures.

If for some reason the development project manager will not manage the implementation phase, the person selected to manage it must become involved with the project early and deeply. By the time implementation begins, he must know as much about the technology of the new process and its requirements for implementation as any person on earth, and more if possible.

When the person who will manage the implementation is selected, the decision should be formally announced in a memorandum from top management stating their unequivocal support for him, his team, and the project. This memo must explain that he has the authority to require, not request, the support and cooperation of the other departments involved.

If top management cannot bring themselves to grant this authority and publicly state their commitment to the success of the project, they should ask themselves, why? Do they wish to hide in the woods until the battle is over and then, depending upon the outcome, emerge to lead the victory parade or bayonet the wounded?

There must be a total and publicly stated top-management commitment to implementing new technology. If the selection and development process was correctly conducted and managed, as explained in the previous chapters, upper management has had plenty of opportunity to have all questions answered and doubts eliminated. By the time the innovation is ready for introduction into production, no doubts should remain, and top management should make it clear to everyone involved that this thing is going to be made to work.

Some managers, when confronted with difficulties at the interface between two of their departments, evade their responsibility to decide by acting as referees. In some situations, this technique of allowing the two sides to continue an adversary relationship while postponing a decision may be appropriate, but the implementation of new technology to which the company has already made a substantial commitment of resources is certainly not one of them. Top management must not now evenhandedly mediate an argument about whether the company should have gotten involved with the new technology at all.

The basic premise of this book is that each step in the development and implementation of advanced manufacturing technology should be taken based on hard experimental evidence about the technology and its suitability to the particular application. If this procedure has been followed and the project has succeeded to this point, the process and the equipment have been rigorously tested and proven to work. There should be absolutely no doubt in upper management's mind that the technology can be successfully implemented. They should not now entertain second thoughts, and they certainly should not make those who have successfully managed the project to this point justify it all over again to a hostile manufacturing department. Although it is necessary to understand the shop's concerns about the new technology and do everything possible to answer them, it should be made very clear that they cannot now veto the project.

The same team building which is important in the development phase between the project staff and the equipment manufacturer is just as important during the implementation phase. Now the team must be enlarged to include key first- and second-level manufacturing supervisors.

This team building requires reassignment of shop supervision so that those who must eventually supervise the operation of the new process will become deeply involved in its implementation. The production foreman should be relieved of some duties so that he can do this. It is irrational to expect a harried supervisor to become involved with a new technology in addition to his other normal duties, without seeing it as just another headache to be evaded as much as possible. Positive involvement of first-line shop management is an important requirement for prompt, successful implementation. It is very difficult to transfer a new technology unless there is someone to receive it.[1] He must be available. It cannot be given to him if he is at a union hearing, searching for lost tools, working on a United Fund drive, or expediting routine production.

Top management's unequivocal commitment to implementing the new technology is best demonstrated by their willingness to inconvenience themselves by changing manufacturing personnel assignments to facilitate it. The same foreman who will see the new technology as an intolerable disruption if it is added to his other responsibilities will strive to make it work if top management relieves him of most of his routine duties, so he can truly become a member of the implementation team.

As in the development phase of the program, if manufacturing middle management is allowed to select the people to join the project team, they will usually pick those they can most readily spare. These are their least skilled, least dedicated people. Such supervisors are usually those least respected by those working for them, and therefore their assignment to the project will signal, not management's concern for the project, but just the opposite, to labor.

The best first-line supervisors, those that shop management thinks it can least spare from the day-to-day battle of the production schedule, are the ones that must be spared. Implementing a new manufacturing technology is much more difficult than merely shoving parts through the old, and much more important to the company's future. Therefore, the best, not the most easily spared, people must be assigned to the project.

This specific assignment to the project team is very important. If the manufacturing people are merely told to cooperate and help when time is available, they will feel no involvement with the project and no responsibility for its final success or failure. Because this will signal management's lack of commitment, they will carefully avoid any commitment themselves.

If the new process is dependent on advanced equipment, which will be a significant technical step for the maintenance department, a maintenance supervisor should also be assigned to the project team. He should be selected for the same reasons as the manufacturing foreman. These members should be named to the project team before the final acceptance trials of the equipment begin.

If the assignment of the manufacturing personnel to the project team with the project staff and the equipment vendor's people is correctly done, a team effort results and one of the most common causes of implementation failures is avoided. This is the three-sided war between engineering, manufacturing, and the equipment vendor that is usually the final phase of a failed advanced manufacturing technology project.

THE T-53 COMPRESSOR ROTOR IMPLEMENTATION

The T-53 gas turbine engine is one of the most common aircraft engines in the world (Figure 6.2). It is primarily a helicopter engine, and about 20,000 have been built. Originally introduced in the fifties as an 800-horsepower engine, improvements had increased its rated power to almost 1400 horsepower in the L-13 model, introduced in 1968. All models have a five-stage axial compressor and a single-stage titanium centrifugal impeller to compress the air which, when mixed with fuel and ignited, produces the train of hot gases which drive the engine's turbine stages (Figure 6.3). The compressor rotates at approximately 26,000 revolutions per minute in normal operation.

In earlier engine models, stages two through five of the compressor were aluminum disks with broached slots into which the compressor blades were fitted, separated by steel spacers. These were bolted together to make the compressor assembly.

In the early 1960s, an electron beam welding machine, then a very advanced process, had been acquired and installed in the manufacturing technology laboratory. A number of projects to develop applications of the process were conducted. The most complex of these was the development of an improved compressor whose manufacture was based on the ability of electron beam welding to make extremely high quality welds with essentially no distortion.

Low distortion was necessary because the requirements of the blade slot broaching process made the finish machining of the individual disks before welding mandatory (Figure 7.1). The assembly after welding had to meet very close tolerances as well as a final dynamic balance requirement (Figure 7.2).

The development project was very carefully conducted and was successful. It developed all of the technology necessary to manufacture and

Figure 7.1 T53 Rotor Components

inspect the component and verified it by making about 60 assemblies which were stringently tested. In addition, a very detailed evaluation of the welds by various metallurgical and mechanical testing techniques was conducted. As described in Chapter 6, a great deal of work to develop the nondestructive inspection technology was also done.

Yet, when the L-13 model for which it was developed was introduced,

Figure 7.2 Welded T53 rotor.

the new compressor was not included. The aluminum disk and steel spacer bolted assembly was used because the new rotor was thought to require too many large steps in both turbine engine design and advanced manufacturing technology implementation.

This position was reconsidered when it was found that the operating conditions in the new, higher-powered engine were more than the old compressor could survive. Failures began to occur in the field. These made the risks of implementing the new rotor much more acceptable. They also made the implementation very urgent.

Not only were all new engines to have the improved rotor, but all engines already in the field were to be retrofitted. There were those, both in the company and in the government agencies involved, who thought that replacing a bolted rotor made by the conventional technology, no matter what its problems, with the new, welded one was a mistake. There was, however, no alternative, and the new design was backed by the results of a well-planned and meticulously conducted development project. Again and again, people who thought such an advanced design could never operate dependably in an aircraft engine were unable to object once they had been briefed on the development project and its results, except for one thing. They always expressed grave doubts that the manufacturing process could be made to work and kept in control in the real world of the production shop.

The implementation project began with a meeting at which top management announced that the project had to succeed and that no effort would be spared to assure that it succeeded. This was the single most important step in assuring successful implementation.

First production was to be started in a then new plant, since closed, in the South. When this implementation was complete, production would begin in the company's main plant. Many years later, rotor manufacture was started in the plant of a licensee, a major Japanese manufacturing firm. The same person managed all three.

Although there was an unequivocal top-management commitment to successful implementation in all three places, by far the most difficult was the one in the main plant. Yet, here it represented the smallest technical step, because the old rotor had been made there for many years and at least some of the new processes required were already in use, if on a limited basis.

Why should the first implementation in a new plant with much less skilled labor have been not much more difficult than a production start-up years later in a Japanese company, famous for its skill at adopting new technology? To begin with, no one in the new plant felt threatened by the new technology or by those who had developed it and were now going to implement it. The success of the project was seen as essential to the continued existence of the plant, and when the project engineer arrived to begin implementation, he was viewed by the management, staff, and labor of the plant as someone who was going to give them something valuable. Anyone who had reservations about the project carefully kept them to themselves.

The engineer was very junior in the main plant,where he normally worked. So junior, in fact, that one of his biggest administrative problems was not receiving mail. In the rigid hierarchical structure that existed, he was not senior enough to be included in the distribution of memos from the exalted managers in manufacturing operations. Memos intended for him were therefore addressed to his boss, with all of the obvious problems of missed or late delivery.

In the new plant they had no time for such frivolities. The factory manager and director of engineering met with him daily to assure that everything necessary to success was done promptly and well. The only personnel problems he had were the result of overzealous people carrying out misunderstood instructions.

The other factor that greatly helped this implementation was that the first-line manufacturing supervisor and the process control and nondestructive testing people who would eventually operate the process were brought to the development laboratory at the main plant, where they were not only shown what was to be done but were given a chance to participate in the manufacture of the last lot of experimental rotors. This was an important step in bridging the gap between the departments involved.

Once the new plant was successfully producing rotors, work began to implement the innovation in the main plant and the real problems started. The line between engineering and manufacturing was then very sharply drawn there, and it was not only impossible to get manufacturing management to assign anyone full-time to the implementation team, they would not even walk across the street to the laboratory to look at or discuss the requirements of the titanium rotor process.

They established their own implementation team from which, until top management finally interceded, everyone who had been involved in the rotor's development and original implementation was excluded. They could not argue the viability of the process now that several thousand rotors had been made and it had been approved by both the Defense Department and the Federal Aviation Agency. But, they could try to prove that it could be made much more cheaply and easily and would have been, if they had developed the manufacturing process themselves. This is the essence of the not-invented-here syndrome.

Every step in the proven manufacturing process was contested. The preweld cleaning process which was critical to controlling weld quality was not really necessary, would damage the parts, was too expensive, and in any case, could not be done as specified because the union contract would not permit it.

The specialized techniques for deburring the broached slots and shot-peening and anodizing the finished assemblies, and the special controls of cutting fluids and degreasing procedures which were necessary to prevent chloride contamination of the titanium, were either inconvenient, not necesssary, or impossible.

The old bolted rotor never required the specialized nondestructive tests needed to assure the integrity of the new one, therefore these tests were unnecessary and would not be done.

Although the manufacturing process had been proven by its success in the new plant and was making scores of rotors to the satisfaction of everyone concerned, even the most conservative agencies of the U.S. government, every step had to be proven all over again to people who would repeatedly raise questions and then refuse to listen to the answers.

If those now long retired manufacturing supervisors and managers were not consciously trying to injure the company, why did they behave this way?

They had all risen through the ranks, had been with the company for many years, and were mostly 20 years older than the project engineer. Those at the new plant did not perceive him as a threat because they knew their reward for a successful implementation would be his return to the main plant. At the main plant they had no such hope. While top management's support at the new plant had been unequivocal, here, because that success had reduced the urgency, and the resistance of manufacturing management was so much greater, they hedged and never did state who, exactly, was in charge.

Manufacturing management felt that their positions, and the job skills upon which they were based, were threatened, and they responded to that threat. Eventually, the process was implemented, but not because anyone led them or convinced them that doing the job in an efficient professional manner was best for both them and the company. Eventually, when the implementation was months behind schedule, top management was forced to intervene and, by brute force, make them do what was necessary. It was unpleasant and very wasteful of both company and personnel resources.

A few years ago, the manufacture of the T-53 rotor was included in a license sold to a large Japanese manufacturing company. The problems were those of explaining the technology and making minor modifications to procedures and techniques required by the small differences in equipment models one always encounters when moving from shop to shop. The language problem, of course, is the interface at which everything can fail if extraordinary care is not taken, but conveying the necessary information and insight about the process was much easier than conveying it across the political gulf that had existed the main plant.

Contrary to popular myth, the Japanese have no magic method for leaping technical hurdles in a single bound, and the advance in technology the rotor represented was about the same as it had been for the main plant and was much less than it had been in the new one. Some of the processes involved were in operation, on a limited basis. Others were started from scratch. Everyone pitched in and did the job, and that was that.

Why was this so difficult in the main plant where it should have been easiest, and why is it so often impossible in American industry? What is the basis of this sullen resistance to technological improvement?

RESISTANCE TO ADVANCED MANUFACTURING TECHNOLOGY IMPLEMENTATION

Innovation destroys capital.[1] The implementation of new manufacturing technology not only makes the old process equipment obsolete, but it makes the personal capital of the company's employees obsolete as well. The skills with the old technology acquired over years are no longer useful. The techniques of problem solving and the way that responsibilities and tasks have been defined must be changed. The insight gained and the self-esteem derived from possessing that insight are threatened. People from other departments, with different skills, now know more about what must be done by the department than those who have been working there for years.

These fears are not groundless, even if not consciously understood. Personal insight into the company's operations gained through years of hard work is usually the only income-producing asset that shop labor and first- and second-level management themselves own. If the new technology is seen as destroying the value of this asset without replacing it with something of more value in the employement market place, implementation will be tenaciously resisted.

An engineer with one or more college degrees, numerous outside professional contacts, and solid experience in the new technology usually cannot comprehend the career situation of a first-line manufacturing supervisor or middle manager. This person often does not have a college degree, and because his job does not involve contact with other companies or any professional society activities, he has no business contacts outside of the company. Often, he joined the company as an hourly worker, and through dedication, talent, hard work, and perhaps some animal cunning, rose to his position as a first- or second-level supervisor. Everything he knows he learned from the company's existing operations. He is usually an intensely loyal, long-term employee who, although he may continually complain about the company, would never dream of leaving and has built his life upon his job with the company rather than upon the more general concept of a career in a profession which might, over a lifetime, involve several companies.

The preservation of the company's capital and the relevance of his skills to the company's operations are his greatest concerns. If his position is threatened, he may see his family's future and his own life as being threatened.

Into this person's department comes a new technology and one or more engineers who do not have the slightest idea of what they represent to him. They either patronize him or ignore him, while strongly implying that everything he has learned over many years is obsolete. They own the future, and at this moment it is being slid into position on his production floor and connected to his utilities.

The production supervisor and middle manager may not know anything about the new technology, but they invariably know plenty about the union contract, lower-level company politics, and exactly what to tell manufacturing management in order to discredit the invaders.

That the worst resistance to the introduction of new manufacturing technology comes from labor is, in my experience, a myth. First-level supervision and middle management resist the most; they see themselves as having the most to lose by its introduction. That labor is usually blamed is probably the result of their political adroitness.

In order to impede the implementation of new technology, while appearing to be protecting the company's best interests, certain common techniques are usually used to the maximum extent that top management tolerates them or project management is not wily enough to discredit them.

The first of these is the argument of the perfect alternative, which in this case is not some undefined ideal but the old technology, if both it and the new produce the same product. The old, broken down machine, which has been blamed for every missed delivery schedule and rejected product for years, miraculously becomes the ultimate technological marvel when its obsolescence will make the supervisor's own skills obsolete. If the new technology is to be quantitatively compared to the old, production department personnel must never be allowed to provide the data on the process upon which they may feel their careers are based. A complete, objective analysis of both processes by the same impartial method must be done.

This resistance is best counteracted by top management making it very clear that the old technology is going out the door on a specific date and any attempts to preserve it will not be tolerated. If project management is not blessed with such a top management, they had better be prepared for a prolonged battle of attrition, in which they will have to quibble and argue and deal with ten political problems for every technical one.

In such a battle it is best to force the argument onto the most objective level possible, and place top management in a position where it must choose between supporting the new technology or admitting that under its administration, the company was incapable of selecting, developing, and implementing it.

Another technique for impeding progress, while claiming to be guarding the company's interest, is to reduce the risks associated with operating the new process by assigning the worst input material and the worst labor to it. This is justified by saying that until the new technique has been "proven," it is best not to risk any acceptable input material or waste the time of any intelligent, dedicated labor on it. Since most manufacturing processes are not designed to convert garbage into gold, this is an excellent technique for discrediting both the new technology and those who developed it. In a case with which I am familiar, only scrap input material, which the old process could not accept at all, was assigned to a new process. When it managed to salvage 75 percent of this and convert it to useful product, the story was broadcast that the new process could not do better than a 25 percent reject rate. Of course, any process capable of converting three-quarters of the scrap fed to it into acceptable products is going to succeed eventually, as happened in this case. But a great deal of time and effort was wasted because management tolerated this technique for impeding progress. Here, too, only the less skilled, less intelligent, and less dedicated employees were trained on the new system. The day-to-day operation of the old technology, operating in parallel, was too important to be jeopardized by allowing the best employees, no matter how interested or how much they wanted to be involved with the new system, to be trained on it. This technique can only be used if top management has not firmly and publicly committed itself to the success of the new system.

The new technology must be given the best that the company can provide, both in materials and in people, for it is the key to the company's survival. If top management still has doubts about the production suitability of the new technology, one can only ask where have they been, and who has been approving the expenditure of resources, throughout the development process. The suitability of the new technology to the company's manufacturing operation must, in a well-managed project in a well-managed company, have been proven before production implementation is begun. If top management has allowed the project to reach this point without *knowing* that it will work, then the company has bad trouble that has nothing to do with advanced manufacturing technology.

Even in cases where top-management support is public and solid, certain techniques for appearing to support the new technology, while in fact resisting it, occur. The most common form of this is to make "improvements" either by ignoring stated requirements or procedures or by changing them. This can be, in its more subtle forms, a difficult problem to confront, because new processes often are developed with redundant controls and conservative margins of safety which can be relaxed somewhat without difficulties occurring. Most production personnel are used to operating closer to the edge of disaster (if not over it) than are development departments. As with all interface problems, this can only be solved by one group having authority across the organizational interface. In this case, until the new process has been up and running for sufficient time to allow a rational reevaluation of requirements, again based on evidence, no relaxation should be permitted. Therefore, the development department, which still knows the most about the new technology, must continue to have complete technical authority until implementation is complete. It is extremely irritating for someone who has developed and set up a process to hear from top management that it does not work, and then to find that it has in effect been modified out of existence, and what does not work is not the process, but the weird mutation produced by unauthorized changes.

Labor and labor unions are alleged to be the worst enemy that the implementation of advanced manufacturing technology has. This is supposed to be motivated by the fear that they will be displaced by the technology whose purpose most commonly is to gain a competitive advantage by reducing unit labor costs. A union will only resist a new technology to the extent management permits them or forces them to resist.

Most innovations which aim at reducing labor costs do so by eliminating the least desirable of jobs. Those that are most easily automated are usually dull, repetitive, and unpleasant tasks. If these are automated, and the company's competitive position is improved, the overall result is often substantial growth in the company's volume of business. This means that, overall, jobs are not lost so much as undesirable jobs are replaced with more interesting work requiring higher skills.

If corporate policy is to replace rather than upgrade its employees into these better jobs which technological improvements create, of course there will be great resistance to these projects. If the company tries to prevent resistance by secrecy and intentional misinformation about advanced manufacturing technology projects, they will only make things worse. The best technique is to honestly inform the employees about the innovation before implementation begins. This should be through a meeting or series of meetings at which both the technology and its impact on them is honestly explained to the labor force. Although these meetings should be conducted by the project manager, an important member of top management should be present to emphasize the importance of the project and to answer questions about its effects on corporate policy.

If such a person is not present, and questions about personnel policies arise, as they almost surely will, the inability of the project manager to answer them and make commitments will undermine the effort to gain labor support. If their support is intelligently and honestly solicited and their reasonable questions reasonably answered, American labor is intelligent enough, especially with the carcasses of several once important industries before them, to know that no corporation is immortal and that 100 jobs in a healthy company are better than 1000 in a dead one.

The worse a company has been run in the past, the more difficult it will be to introduce a new technology. Poor management not only causes bitter adversary relationships, but also causes and tolerates overstaffing and overly rigid work rules and job classifications. If the old process was overstaffed or was a source of continuing overtime to meet delivery commitments, the new technology will be much more of a threat to labor than it would be if the old process had been well managed. If the introduction of the new technology must correct long-term management errors, in addition to overcoming the natural resistance to change, it will be even more tenaciously resisted.

The classic case of not modifying the work rules to fit the new technology, and thus not gaining the maximum advantage from it, is the fireman on an electric locomotive. There is a much worse situation possible in a manufacturing shop than retaining someone whose job no longer exists; operations that are absolutely necessary to the success of the new process cannot be implemented because of an existing labor agreement.

For example, the metallurgy of the final weld on the power turbine shaft assembly (see chapter 5) leads to its being very brittle and thus delicate as it comes from the welding machine. If it were to be transferred in this condition from the welding shop to the heat treating department, jolting during this transfer could cause cracking. Once the welds have been heat treated, this risk is eliminated along with the brittle condition. Under the existing union contract, at the time of implementation, all heat treating was done in the heat treating department by its personnel. Welding department personnel were not permitted to do it.

The technical logic of the situation required that the new, automated heat treating facility, built specifically for this job, be installed near the welding machine so that the welder could gently remove the parts and place them directly in the heat treating machine. All sorts of objections were raised in the name of the union, as were all sorts of silly compromises, such as having the welder contact his foreman who would contact the heat treating foreman, who would send a heat treater to the welding area to take the part from the welder's hands and place it in the heat treating machine. Often, such things are accepted as givens, and the value of the new process to the company is substantially reduced. This should not be tolerated, because it threatens the company's future competitiveness. In this case, the requirements of the technology were explained to the union, and they saw that this was not just another plot to erode their position. A sensible modification of the work rules was agreed.

The T-53 titanium rotor also provided an example of this problem. Parts for electron beam welding were normally cleaned by the welding department personnel who then assembled them. The separation of these functions makes no sense because all accountability for the welded assembly is lost. If responsibility for cleaning and welding are divided, the welder blames the person who did the cleaning for all rejects, and no one is responsible for the quality of the welded assembly. Prior to the titanium rotor, this cleaning operation always consisted of mechanical cleaning with wire brushes or such, and wiping with solvents. The rotor required a procedure based upon an acid etch.

Under the standard job description, all chemical operations such as this were done in the plating room by its personnel. This made no technical sense for the rotor, both because of the divided responsibility that would result, and because of the risk of recontamination during transit between departments. The best technique, from the point of view of doing the job right, was to set up a facility for chemically cleaning the parts adjacent to the welding area. Again, there was all sorts of bureaucratic bickering, and again, when the technical basis for the change was explained to the union, they agreed.

One should never accept the opinion of the personnel department, or of a shop supervisor, that the technically superior solution cannot be implemented because the union will never agree. No matter how militant the union, at least their motivations are rational, which makes them easier to comprehend and deal with than the subconscious compulsions of some lower-level corporate bureaucrats.

The actual shop labor, as opposed to the union officials, often adjust more quickly to the new technology and come to be its earliest and most enthusiastic supporters. Perhaps this is because of the obvious discomfort it so often causes the foremen. Shop labor usually see it as an opportunity to improve and upgrade their job skills. The foreman, especially if he is permitted to evade involvement in the project, often sees this, not as an upgrading of his employees' skills, but as a denigration of his own. Such a foreman will only assign his least dependable and untalented people to the new job, and he will assure that any instructions he is asked to transmit are hopelessly garbled. If, however, top management is frequently seen on the production floor, asking elementary questions and freely admitting their own lack of omniscience, it is much easier for first- and second-level supervision to also behave in the company's best interests.

Every question asked of the project staff must be correctly answered. This is another reason why a professionally planned and conducted development project is an absolute requirement for successful implementation. Whether the question appears brilliant or stupid, and whether it is asked by the president of the company or by the sweeper, it must be answered. If the answer is unknown, this should be admitted. Then the answer should either be found and transmitted to the questioner, or someone who can answer it should be put in direct touch with him. People who act as though a question is beneath them do so because it is above them. The success of the project is dependent on those who develop and implement it gaining profound insight into the technology. Implementation is the transmittal of that insight to the shop personnel who will operate the technology.

One of the most pathetic shows one can watch in a production shop where a new piece of equipment is being installed is a project manager, like the one in the case study in Chapter 6, disdaining to answer questions that are beneath his dignity about a system that he cannot turn on, turn off, or operate himself. No matter how haughty his manner, no matter how silky his evasions, that system is in very deep trouble, if not doomed.

When someone in the shop asks a central technical question about the technology, as they certainly will, the very best way to answer it is, "Here, let me show you . . . ."

ENDING THE DEVELOPMENT PROJECT

Striving for perfection in the new manufacturing process can become such an obsession to those who conduct the development project that they may not stop when they should. Once the final testing of the process and the equipment has been rigorously and successfully conducted; and all of the prestated requirements have been fully met; and a proven, production-suitable manufacturing process and its equipment are ready for implementation, stop developing it.

Of course, there are further improvements that can be made and additional features that can be installed. If these are required for the project to meet its stated objectives or for the innovation to be successful in production, then the project is not completed. But if these improvements are not required to meet the requirements of either the project or the production shop, if they are nice-to-haves rather than got-to-haves, do not include them.

Because these improvements or accessories were not part of the development project, neither resources nor preagreed acceptance criteria to evaluate them exist. Do not degrade the performance and complicate implementation by adding unnecessary and insufficiently tested modifications. Leave them for follow-on projects.

The process and the equipment which were completely and successfully tested at the end of the development project are those that must be implemented in production. Sometimes, when the successfully tested process arrives, with public backing of top management to its success, the not-invented-here syndrome appears in the form of suggestions for improvements from those who feel threatened by the innovation. Corporate politicians much favor the technique of getting some small modification made, they really do not care what, at their suggestion. Then, if the system works, they point to this trivial and sometimes ludicrous thingamajig as the only reason that the project succeeded. If the system fails, sometimes because of their constant behind-the-scenes resistance, they point to the thingamajig as proof that they did their very best to support it.

It is sometimes very difficult to get the project staff to stop tinkering, particularly if they see the conclusion of the project as a career crisis. The best way to assure that no end to the project is ever reached is to let those conducting it know that a layoff or return to a routine dead-end job will be their reward for completion. Management should now begin discussing ideas for the next project with them, and assure them that as soon as the current project is completed by its successful implementation, the next project will begin and they, because of this success, will be at the center of it.

IMPLEMENTATION APPROACHES

The first step in implementing a new manufacturing technology is to select an overall approach for the implementation. It must be consciously devised. Just as developing a new manufacturing technology is a great deal more than merely buying the equipment, implementing that technology is a great deal more than just installing it on the production floor. Everything that can be done to make the production start-up more efficient must be done.

If the new technology is to perform an operation currently being done by an older method, parallel implementation, in which the old process continues to run while the new is being brought on line, can be used. This technique, where possible, reduces the disruption of the production schedule, but it does involve space and work flow problems that must be considered. This scheme can cause real problems if it is used without a firm commitment to the new technology by top management. Everyone involved must know that this is not a contest between the old and the new, but that everyone must work toward shutting down the old, less efficient process just as soon as possible.

The alternative to parallel implementation in situations where the new technology replaces an existing process for making the same product is to shut down, tear out the old system over a weekend or holiday, and crank up the new. To be honest, I prefer this approach because its high apparent risk forces everyone to do his job correctly the first time. Someone who always wears a life jacket will never learn to swim. If the development program has been correctly conducted and the new manufacturing process has been selected, debugged, and tested the way any new process should be, then there is nothing to foster communication and cooperation between departments quite like the certain knowledge that they are all going to be on the unemployment line together if they fail. Narrow departmental parochialism and a lack of this "we are all in it together" attitude are the major stumbling blocks to the adoption of advanced manufacturing technology. The other major barrier is lack of nerve on the part of top management. This lack of nerve manifests itself first in an unwillingness to risk anything on the development of new manufacturing techniques, then, when development is grudgingly permitted, in attempts to stay uncommitted long past the point where their commitment is an absolute prerequisite for success. Parallel implementation is a convenient technique for management hedging.

If the new technology truly is an improvement which will provide the company with a substantial competitive advantage by its adoption, and if it is ready for implementation as proved by rigorous testing to logically defined criteria, why go through the pain of entering the future one toe at a time.

If, however, the step from the old technology to the new is a very large one for the particular shop, and a great deal of training of supervision, labor, and maintenance personnel is required, a parallel implementation may be prudent. But management should know the difference between prudence and cowardice. If a parallel implementation scheme is adopted, assuming that the new technology will replace the old and that an increase in production schedule does not require that the old process remain in operation, a definite target date for shutting down the old system should be set by top management and enforced. If not, excuses for keeping the old one, just in case, will assure that it continues to waste company resources that could be much better spent in further improving the new technology.

The most difficult implementations from a management standpoint, in which I have been involved, have been parallel implementations where an unequivocal commitment to success was not required by the circumstances and was therefore not made. This is ironic, because one would expect that the production start-up of a critically needed new part that could only be made by a new process involving a number of new operations would be much more difficult. Although the technical complexity of such a project is much greater, the firm top-management commitment to success that its desperate nature generates leads to much less conflict and greatly expedites completion.

THE IMPLEMENTATION PLAN

The development project is completed. The new manufacturing process and its equipment have been developed, debugged, tested, and documented. Like everything else in the project, the implementation should be carefully planned, taking into account the specific requirements of both the existing shop and the new technology, so that they can be meshed in the most efficient way. As always, every task required for success must be made the responsibility of one specific person who knows what he is to do, why he is doing it, and when it must be completed.

The various tasks must, of course, be assigned to those on the project team best able to perform them. If no production supervisor has been assigned to the team, there are certain critical tasks which will either not be done or be done badly, because they require his knowledge of the production shop and his line authority there. These are the selection and assignment of the best people to the new process, the negotiation of any required job description or labor agreement changes, and most important, communication about the project to labor. All of these functions are best done by a competent supervisor who has the respect of those who work for him.

The problem of providing technical support to second- and third-shift operations by the project staff must be addressed. Often, engineers do not work nights except very grudgingly, and therefore the later shifts are provided insufficient technical support. Particularly because night shifts are sometimes used as penal colonies for first-line supervisors who are being punished by manufacturing middle management, the opportunity for trouble if insufficient technical support is provided is very great. The project staff has to address this problem and work out an equitable schedule to assure that these shifts are adequately supported.

Technical support must be provided, not only to the start-up of the new equipment and advanced technology operations but to the entire manufacturing process of which they are but a part. If the development project has been competently planned and conducted, the implementation of the advanced technology operations should be relatively easy, particularly because everyone's attention will be directed to them. But often the worst problems are not with these operations at all, but with older, in-place operations integrated into the new manufacturing process and at the interface between them and the new technology.

The implementation of the power turbine shaft (see Chapter 5) met its greatest difficulties not with the electron beam welding or inertia welding or advanced ultrasonic inspection operations, but in the lathe and grinding operations on the detail parts prior to assembly for welding. The machine shop people had to consistently meet tolerances tighter than had traditionally been required, although well within the capabilities of the equipment and processes involved. These narrowed tolerances, and the fact that they were not negotiable, were the greatest single problem in the implementation. The machine shop was so used to having tolerances overspecified and then relaxed as reality required that it was very difficult to convince them that the engineering drawing limits truly were the minimum acceptable; they were not just intended as goals to be met if convenient. Repeatedly, parts were delivered to the welding operations and sent back for remachining when they did not meet the preweld dimensional requirements. One can imagine the level of civility with which these repeated rejections were discussed. Many of these problems could have been reduced if positive steps had been taken to inform machine shop management about the importance of, and reasons for, these narrowed tolerances, before production began.

EQUIPMENT REPAIR AND MAINTENANCE

Many advanced manufacturing innovations fail on the production floor for no better reason than that the maintenance department either cannot or will not keep the equipment upon which the new process depends running. The question of what to do if they cannot service the equipment is by far the easier of the two, so we will discuss the question of why they will not service it first.

Maintenance departments are nonproductive overhead operations. They are also legitimate and essential costs of staying in business. They are as critical to the long-term survival of a manufacturing company as a fire department is to the survival of a city.

Unfortunately, this fact can be ignored, and often is, by managers looking to improve their own short-term profit appearance at the expense of the company's long-term health. When the economy turns downward, the first department to be cut is usually maintenance. A good argument could be made that during a recession, when the efficiency and flexibility of the company's manufacturing operations are critical, the department responsible for the service and repair of its equipment should be strengthened, not weakened. But it never is. In those tough times the argument could be made that the paper shuffling bureaucracy which keeps the company from reacting quickly and logically to changed market conditions should be ruthlessly cut long before the condition of the company's plant and facilities are allowed to deteriorate. But it never is.

Maintenance departments are usually the first to be cut, whenever cuts must be made, unless, of course, those who work in them take steps to protect themselves. In many companies, the overriding concern of all of the employees of the maintenance department is to protect their jobs. They accomplish this by not doing them.

Experience has taught them that the only way they can be safe from capricious cost cutting is to keep a large backlog of broken, desperately needed equipment. If they were to exert themselves, in the company's best interests, and eliminated this backlog, their reward would be the unemployment office. These departments are almost always the most militantly union, the most rigid in following work rules, the least concerned about the health of the company, and the ones most devoted to featherbedding. It may take weeks to get a maintenance man to look at a broken machine, but when he finally arrives, he will not be alone. He will have a helper. Maintenance men, like teenage girls, never go anyplace alone.

Such maintenance departments, like everything else in the company, are the fault of management and can only be corrected by management. If this situation, and the deep mistrust of management's motives and intelligence which caused it, has festered for very long, it will take extraordinary effort to reverse it. Management must sit down with the maintenance personnel and somehow convince them of their own good faith, so that negotiation of some sensible solution can begin. This is an essential and a difficult situation to solve. If it is not addressed, millions of dollars of the company's money will be wasted because trivial repairs are turned into major problems by people who are convinced that their best defense against their own management is to make sure that nothing works very well, if at all.

Some union work rules present great difficulties in the servicing and repair of complex advanced technology equipment. There is a government installation at which the maintenance people routinely change jobs every few months to assure "fairness" in job difficulty and to equalize skills. This makes the operation of complex equipment just about impossible, because no one stays long enough on any one job to develop any of the insight needed to do it.

The most generally found and most difficult work rule problem, however, is the one caused by trying to install, repair, and maintain complex modern equipment while dividing the individual tasks between trades that were defined 50 or 100 years ago.

On a certain model of electron beam welder, there is an air-actuated, electrically controlled valve whose stem seal is prone to wear. When it begins to leak, the machine cannot operate. Working alone, a competent mechanic can remove the necessary casings, disconnect, remove, and overhaul the valve, and then reinstall it in about four hours.

This same job, if done under the classic union contract definition of the duties of the various trades, requires a machine repairman, a millwright, an electrician, a plumber, and a rigger (each with his helper). In most shops, it is not possible to round up all of these people and have them and their tools available in four days.

No matter how suitable the new technology, or how well the application is developed and how great the competitive advantage it will provide, and no matter how much top management wants it to work, if the only people who are permitted to fix the equipment upon which its success depends will not fix it, the project fails. The machinery goes down and it stays down. This basic reality of advanced manufacturing technology development and implementation must be understood and definite steps taken to assure that when the system goes down, someone with the skills, the tools, the spare parts, and the will to fix it is available to fix it.

Training of Maintenance Personnel

Usually the equipment manufacturer includes with the equipment a training course for both operators and maintenance personnel. The maintenance courses are usually useless. A 20-hour course on the maintenance of an essentially standard machine may be adequate for someone who is completely familiar with similar machines and has only to learn about the special features of this machine.

If, as is almost always the case with advanced manufacturing technology, the equipment is the first of its type any of those involved have ever seen, the course may need to be several weeks long. The time required should be based on what must be taught and to whom. It should not be based on some arbitrary estimate of the maximum time that can be permitted, either by the vendor or the customer.

The course must be conducted off-site, usually in the vendor's facility. If that is not possible, rent a hotel room at least 25 miles from the customer's shop. Attempting to conduct an in-depth course about a new technology within 25 miles of most companies, with their understaffed maintenance departments, is a waste of time. Do not even try. The interruptions, while supervisors and manufacturing management pull people out to fix constant, but always urgent, problems, will turn the course into a shambles.

Those who attend the course must be those who will work on the equipment. Do not send someone who is about to retire away to school, as a going-away present.

The best time to schedule this course is before the new equipment is shipped, while it is undergoing final trials at the vendor's facilities. Once it is accepted for shipment, the customer's maintenance personnel can help with the disassembly. Then, when the equipment arrives on the production floor, the maintenance department can be truly involved in the installation and continue to learn about it during start-up.

The purpose of the course is to provide the insight into the operation of the equipment necessary to diagnose faults and to repair them. In many cases, these courses fail to provide even the basic understanding needed to perform preventive maintenance operations without doing damage. This is because the assumptions about the basic skill level of those taking the course are wrong in the particular case, or the course is planned and taught by those who designed the machine, not those who know how to service it. Long explanations of the operating system logic of a computer numerical controller are of much less use than a detailed explanation of the output of the diagnostic program and which circuit board to change when a given symptom occurs. The course must be focused on training people to repair equipment, not design it.

The course should be taught from the drawings and service manuals of the actual system, so that the adequacy of these documents for their intended purpose is determined. Often the vendor will provide impressive-looking manuals, 90 percent of which are later found to be information so general as to be useless, or describing older systems and thus not applicable to the actual system supplied. Perhaps the manuals are so complicated and poorly written that the customer's maintenance people cannot understand them. Early discovery of these problems is another reason for holding the course prior to the shipment of the equipment.

A member of the project staff who is completely familiar with the equipment should take the course along with the maintenance personnel. This should be the person with the most insight, who needs the course least. He should review the manuals to assure their completeness, act as a referee in any disputes between the customer's personnel and the vendor people teaching the course, and also be able later, if someone pleads ignorance, to know exactly what was covered in the course. If he feels the course is inadequate, he should immediately report it to the project manager, who should take action to assure that the vendor corrects the situation.

Spare Parts and Special Tools

No one wants to have money tied up in spare parts inventories, neither the company which owns and operates the equipment nor the company that sold it to them. Yet, the unavailability of a relatively cheap part will not only shut down a million-dollar system, it may destroy the credibility of that system by keeping it from meeting a production schedule. The new system must only disrupt the company's operations a few times before the plan to get rid of the old, expensive process will be abandoned. Rather than concentrating on getting the most from the new process, effort will then shift to keeping the old, "dependable" process running. Often the only reason for the apparent difference in dependability of the new and the old technologies is that the needed spare parts are on hand when the old system breaks down.

Vendors' spare parts lists, particularly for new and specialized systems, are often little more than guesses. When new systems begin to run in production, it is often found that parts which were assumed to be essentially permanent actually have a limited life and must be considered maintenance items. When this happens, this information must be transmitted back to both the customer's purchasing department and the vendor's spare parts sales department, so that these parts can be stocked. If this is not done, a part may only be available as a special-order item from its original manufacturer, a process that may take months. When any part has failed or needed overhaul twice, it should certainly be considered a required spare and definite steps taken to assure that an extra is available, on-site, at all times.

This is a function which most commonly fails at department interfaces. The project staff feels that the provision of spare parts is the responsibility of the purchasing and maintenance departments. The purchasing department replies that they bought the spare parts listed by the vendor. The maintenance department says that when the first one broke, they tried to order both a replacement and a spare, but someone in accounting refused to authorize the purchase of more than one. Now a second replacement is needed, and the manufacturer is quoting ten weeks for delivery. Meanwhile, the system is down.

When this happens, the project manager must intervene to get something done, whether he feels it is his responsibility or not. Often, he will get the equipment vendor to cannibalize his own equipment or a system being built for someone else, and the production system will be repaired.

Unfortunately, like many informal quick fixes, this does nothing to solve the underlying organizational problem, and three months later, when the same part once more fails, again no spare part is available on-site. A spare parts system must be put in place to assure that needed spares are always available. This is an essential function which, like many others, the project manager cannot do by suggestion and requests. He must have the authority to order it done.

Once spare parts are purchased, some system for assuring that they will be available when needed should be instituted. A spare that was used to fix some other machine two weeks ago, or is locked in the toolbox of someone on vacation, is useless. This problem of things disppearing into people's toolboxes is particularly troublesome with manuals and special tools. It is the responsibility of the supervision of the maintenance department, but the project manager must make sure that they deal with it.

He may also have to deal with the problem of financial people who arbitrarily cut items off of every requisition that crosses their desk, without having the foggiest notion of what the items are and why they are needed. If all reasoning fails, a letter with a large distribution, explaining how by saving the company $90 last month the specific person named caused the loss of $9000 per day through the unavailability of a critical spare part, may be necessary. Assuring that those who take credit for short-term savings also get credit for the long-term losses they cause is excellent management at all levels.

Preventive Maintenance

Once the system is running, preventive maintenance and routine servicing of the equipment must be assured. Often, the old equipment has been serviced on the run-it-until-it-breaks system, and the concept of preventive maintenance is no longer understood by manufacturing management. If the old equipment was simple and primitive, and everyone knew exactly what to fix every time it stopped, they will not like a new system which, if not cared for, is liable to go down and stay down because no one can yet diagnose all that can be wrong with it.

Everything possible to keep the new system running should be done, and one of the most important is to perform all scheduled maintenance on time and correctly. It is infuriating for the development staff to have the new system running flawlessly at the end of the day, and arrive in the morning to find it down because someone botched a simple maintenance operation performed at night. If, as is often the case, preventive maintenance is performed on the night shift, technical support and training, not only to the maintenance labor but to their supervision, must be provided. If this is not done, and only a cursory understanding of the equipment is provided, the chances for real problems are very great. Someone from the vendor's service department or from the project staff or both must work with the later shift to assure that they know how to service the equipment. Their supervision must also understand what must be done and why, so that time-saving shortcuts will not be tried. Time-saving shortcuts, devised by people who do not understand equipment, is one of the most common causes of damage.

OPERATOR TRAINING

The training of operators for advanced manufacturing technology equipment ranges from essentially hands-off systems, which require only a spectator who hits a stop button if something odd happens, to someone who sets up and programs complex metal cutting or joining equipment. There are, however, certain principles that are common to all.

The best way for operators to be trained is to first train the person who will supervise those operators, and let him train them. The opposite extreme is a production foreman who does not know, and does not want to know, how to run the equipment his employees must operate. Under such supervision the opportunities for labor to reduce the productivity of the new process and impede its implementation are great. I once encountered a foreman who was scheduling two loads per shift through a welding machine which should have done eight. His operators had convinced him that it was necessary for them to wait, while doing nothing, for three hours for the system to reach operating vacuum. It actually took 20 minutes.

The foreman should be assigned to the project staff toward the end of the development phase and learn to operate the equipment before it arrives on the production floor. Operator training should then be his responsibility, even though the equipment vendor may conduct the actual training course.

The foreman must attend these classes full-time and assure that the instruction is suitable and adequate.

Operator's manuals, like maintenance manuals, often have page after page of irrelevant gibberish. If this was found to be the case during the original training of key personnel at the vendor's facility, it should have been corrected. Many systems are delivered with four-inch-thick manuals containing everything about the system except what one really needs to know to make it work. These systems are found, six months after being started, to have concise instructions developed from necessity taped to their control panels. These usually say if X happens, do Y, and if that does not work, stop and call so-and-so. These instructions usually are developed after someone who had limited insight about the system encountered X and then tried Z, and when that did not work, tried A and B six times, causing serious damage.

The project staff should think their way through these possibilities, prepare such simple instructions, and use them for operator training courses. At least try to anticipate those things which are most likely to stop production or do damage if attempted, and provide simple instructions to prevent them. The best way to prevent these problems is to provide the operators with instructions not only on what and what not to do, but on why.

There is a tendency to attempt to train operators by rote to do only specific procedures and try to keep them from gaining any understanding of the logic behind those procedures. This demeans the operator, no matter how much it inflates the egos of those who favor such policies. No one ever did a worse job because he truly understood what he was doing. If, however, he is given poor or contradictory instructions by those who do not know much themselves, then trouble is sure to result.

Providing in-depth training to the operators, and thus helping them to improve their salable job skills, is an important part of gaining labor's cooperation in implementing advanced manufacturing technology.

PRODUCT QUALITY ASPECTS OF IMPLEMENTATION

There is one type of improved manufacturing technology that is extremely difficult to implement even with top management's complete support, and impossible without it. This is the case where the old technology—either the manufacturing technique, the inspection methods, or the two together—permitted products of poor quality to be shipped to customers. A significant competitive advantage of the new technology is derived, in this case, from the better inspectability and easier control of the manufacturing process, or from an improved inspection technique which discovers defective products before they can injure the company's relationship with its customers.

Implementing such improvements in the real world of a production shop, without demonstrated top-management support, can only be described as a nightmare, because the new technology is not seen as solving major problems, but as causing them. Many American manufacturing departments, because of overly rigid compartmentalization of companies, are totally isolated from the competitive realities of the marketplace. Only when the volume of orders declines and production is reduced is there any contact with the outside world, and this is not viewed as being related to the past performance of the department or the quality of its products, but as the result of outside, uncontrollable forces. A drop in the production schedule and the resulting layoffs are blamed on general economic conditions, or on the political party in power, or on unfair and "predatory" practices by competitors. They are never admitted to be the result of the poor quality or durability of the product the manufacturing facility produced.

If the shop has been making something by an old technology whose defects were only found by the customer, it was therefore trouble-free in production. If now, with the new technology, those defects are found, and manufacturing is forced to take steps to eliminate them, they may try to discredit the new technology.

First, it will be shown that the old manufacturing techniques were dependable and produced no rejects, while the new technique is obviously inferior because defects are now frequently detected. Because products that formerly would have been acceptable must now be reworked or scrapped, the new process is obviously more expensive to operate and disruptive to the production schedule.

This killing of the messenger who tells the unpleasant truth is extremely difficult to explain to a bottom-line-oriented top management, who do not understand the relationship between the design of their company's product, the methods used to make it, the methods by which its quality is controlled, the performance and durability of the product in service, and the competitive position of the company in its market. Rather than being seen as a major competitive opportunity, a new technology which raises the quality of the delivered product by increasing the number of defects found in production will often be quickly discredited and discarded. This is, in the short run, much easier than forcing everyone to cooperate in eliminating the problems that are now potentially solvable because they can be detected.

In order to implement such innovations, top management must have considerable discernment about the details of the business and the interrelationships between them as well as the effects of the innovation. They must also have the maturity to maintain an atmosphere of civility and professionalism while ruthlessly doing what is necessary to assure that the problems are solved, not ignored in the interests of superficial harmony. The ability of a company to implement this type of innovation, and solve the technical and organizational problems that it detects, will have great impact on the company's survival in anything but the least competitive of businesses.

It is very important for the project staff to assure that everyone involved with implementing the innovation knows that the improved control of product quality it provides is an advantage to be exploited, not an impediment to be evaded. This potential improvement in product quality is often very difficult to explain, because no one wants to admit just how bad the old product may have been and would much prefer to blame its failure in the marketplace on factors for which they had no responsibility. This form of cognitive dissonance can kill companies, ironically when the technology which could save them is at hand.

PROCUREMENT OF ADDITIONAL EQUIPMENT

The new system is now up and running, and because of its success, more equipment is needed to exploit the competitive advantage that it is providing to the company. The procurement and implementation of the additional production capacity should be conducted so that the second generation equipment is at least as good as, and hopefully even better than, the original.

Improvements can probably be made in the design of any advanced technology system once it has been run for some time on a production basis. Those who developed and implemented the original system should realize this, so that they will not now become afflicted with the refusal to accept change that they hate in others.

It is assumed that the original system is successful and that the additional equipment will be identical to it in all major aspects. If the new equipment is of a substantially changed nature, so that new central technical questions must be answered, then it is not additional equipment to the original design. In that case, work begins with a feasibility study. Only if the CTQs of the new equipment were answered by the original development program is it a follow-on procurement.

If it is truly additional equipment, certain improvements will probably have been suggested by the experience of running and servicing it in production. In most cases, these will be modifications which improve the durability of components and simplify maintenance. These, of course, will not be made if the corporate procurement system reissues the purchase order and specification for the original machine. Improvements will certainly never be made if someone orders these new systems to this old specification from a different manufacturer because he is the lowest bidder. It is still possible, even at this late date in a successful project, to have the turnkey system of procurement cause a disaster.

If the original system was designed and built based on the insight gained from a pilot plant, many of the modifications will involve redefinition of trade-off choices made during the design process. For example, in the design of the original recuperator laser welder (see Chapter 4), the two work stations were placed fairly close together in order to conserve floor space. This made the servicing of certain work station components difficult if the robot and the other work station were operating, because of the lack of room as the robot arm swung. In the second-generation systems, floor space was not a problem, so it was possible to modify the design to provide adequate space to work.

Decisions about modifications must be made by manufacturing, engineering, and the equipment vendor rationally selecting the best solution. This may cause problems in some companies, because the continued cooperation of these partners in the successful project is seen as collusion to exclude other suppliers from bidding. If the original project was a success, as the decision to order additional equipment would indicate, and if both the development project staff and the manufacturing department are pleased enough with the vendor's performance to want to continue their relationship with him, then, for the good of the company, everything possible, fair or unfair, should be done to assure that he gets the follow-on order.

The company, after expending the resources needed to develop and implement an advanced manufacturing technology manufacturing system, should not accept the technical risk and inconvenience of now trying to implement a different system in production. Fairness, whatever that is, is irrelevant. What matters is what will provide the company with the greatest competitive advantage with the least expenditure of resources in the shortest time.

The expenditure of resources is not calculated by comparing the price quoted by someone who knows how to build a system which works to that of someone who thinks he knows how to build such a system. Low bids are frequently the result of someone's not knowing all the things he must do and hence not including them in his price. The cost of purchasing a system different from the one successfully running must include the training of both manufacturing and maintenance personnel, the increased spare parts inventory required by the loss of commonality, and the need for another development project, including pilot plant, if the systems are at all technically different. It must also include the lost opportunity to work on something else, if the successful project must in effect be repeated. Whether another pilot plant is required must be decided by a detailed evaluation of the technical equivalence of the two systems, based on rationally planned and conducted tests, of course.

If the equipment supplier's performance was unsatisfactory, then an alternative source must be found, and the additional work is certainly justified. It cannot be justified, however, by fear that the company might be criticized for a lack of fairness to all of the equipment suppliers in the universe.

PROJECT COMPLETION

There is a point in the life cycle of a manufacturing process, no matter how advanced, when it is finally implemented, and the project team goes on to other things. This happy time comes when certain criteria are met. The process is routinely being run in a production shop, by production people, under production supervision, dependably meeting a production schedule with a product of acceptable quality, at a competitive price. When these criteria are met, those who nurtured it through its childhood and its perhaps difficult adolescence can consider their project complete.

Problems may arise with the technology requiring the support of those who conceived, developed, and implemented the innovation. Incremental improvements will be made and the products the process makes will undoubtedly evolve. But it is no longer advanced manufacturing technology in that company. It is implemented manufacturing technology. The goal, first mentioned in the statement of purpose of the preliminary experiment, "To implement in production," has been accomplished.

REFERENCE

1. Schumpeter, Joseph as quoted in Hayes, Robert H., and Wheelwright, Steven C. *Restoring Our Competitive Edge—Competing Through Manufacturing,* New York: Wiley, 1984.

BIBLIOGRAPHY

Abernathy, W. J. Competitive decline in U.S. innovation: The management factor. *Research Management,* 34–41 (Sept. 18, 1982).

Anonymous. Miles of laser welds made automatically. *American Machinist,* **127**(11):41–42 (Nov. 1983).

Anonymous. Stimulating industrial innovation. *Research Management,* **18**(6):30–33 (Nov.–Dec. 1984).

Balachandra, R., and Raelin, J. When to kill that R & D project. *Research Management,* 30–33 (July–Aug. 1984).

Becker, R. H. Project selection checklists for research, product development, process development. *Research Management,* 34–36 (Sept. 1980).

Brandt, A. Computer strategy: A philosophy for managing information processing resources. In Goldberg, Robert, and Lorin, Harold (Eds.), *The Economics of Information Processing* (Vol. 1), New York: Wiley, 1982.

Brosilow, R. Coming soon to the production floor . . . . *Welding Design and Fabrication,* 38 (June 1984).

Clifton, David S., and Fyffe, David E. *Project Feasibility Analysis,* New York: Wiley, 1977.

Cohen, H., Keller, S., and Streeter, D. The transfer of technology from research to development. *Research Management,* 11–17 (May 1979).

Cohn, S. F. Adopting innovations in a technology push industry. *Research Management,* 26–31 (Sept. 1981).

Curd, A. J. Avoiding pie in the sky. *Research Management,* 34–41 (Jan. 1982).

Daly, E. B. Organizational philosophies used in software development. In Goldberg, Robert, and Lorin, Harold (Eds.), *The Economics of Information Processing* (Vol. 2), New York: Wiley, 1982.

Davies, Stephen. *The Diffusion of Process Innovations,* Cambridge: Cambridge University Press, 1979.

Dearnley, P. A., and Mayhew, P.J. In favor of system prototypes and their integration into the systems development cycle. *The Computer Journal,* **26**(1):36–48 (Feb. 1983).

DeCotiis, T. A., and Dyer, L. Defining and measuring performance. *Research Management,* 17–22 (Jan. 1979).

Drucker, Peter F. *Management,* New York: Harper & Row, 1973.

Drucker, Peter F. *Managing for Results,* New York: Harper & Row, 1964.

Drucker, Peter F. The innovative company. *Wall Street Journal,* (Feb. 26, 1982).

Etienne, E. C. Interactions between product R & D and process technology. *Research Management,* 22–27 (Jan. 1981).

Ettlie, J. E. The timing and sources of information for the adoption and implementation of production innovations. *I.E.E.E. Transactions of Engineering Management,* **EM 23**(1) (Feb. 1977).

Feinberg, M., and Levenstein, A. Developing and making the most of the slight edge. *Wall Street Journal* (July 9, 1984).

Fernelius, W. C., and Waldo W. H. Role of basic research in industrial innovation. *Research Management,* 36–40 (July 1980).

Florman, Samuel C. *Blaming Technology,* New York: St. Martin's Press, 1981.

Forbes, R. J. *The Conquest of Nature,* New York: Praeger, 1968.

Foster, R. N. Boosting the payoff from R & D. *Research Management,* 22–27 (Jan. 1982).

Foulkes, F., and Hirsh, J. People make robots work. *Harvard Business Review,* 94–102 (Jan.–Feb. 1984).

Frosch, R. A. R & D choices and technology transfer. *Research Management,* 11–14 (May–June 1984).

Gage, J. H. Identification and start-up of a successful robotic arc welding application. *Welding Journal,* **62**(11):19–26 (1983).

Gold, B., Rosegger, G., and Myles, G. *Evaluating Technological Innovations,* Lexington, Mass: Lexington Books, 1980.

Goldberg, Robert T., and Lorin, Harold (Eds.). *The Economics of Information Processing, Vol.1: Management Perspectives,* New York: Wiley, 1982.

Goldberg, Robert T., and Lorin, Harold (Eds.). *The Economics of Information Processing, Vol.2: Operations Programming and Software Models,* New York: Wiley, 1982.

Green, S., Bean, A., and Snavely, B. What happens to ideas. *Research Management,* 21–39 (Nov.–Dec. 1984).

Gremillion, Lee L. Systems development and implementation costs using standardized application systems. In Goldberg, Robert T., and Lorin, Harold (Eds.), *The Economics of Information Processing* (Vol.2), New York: Wiley, 1982.

Hackney, John W. *Control and Management of Capital Projects,* New York: Wiley, 1965.

Harrison, F. L. *Advanced Project Management,* New York: Wiley, 1981.

Hayes, Robert H., and Wheelwright, Steven C. *Restoring Our Competitive Edge—Competing Through Manufacturing,* New York: Wiley, 1984.

Hendrickson, W. What is purchasing's role in the office systems buy? *Purchasing,* 43 (Nov. 1984).

Hersey, Paul, and Blanchard, K. H. *Management of Organizational Behavior,* Englewood Cliffs, N.J.: Prentice-Hall, 1969.

Hill, I. D. The real barriers to innovation. *Research Management,* 6 (Mar.–Apr. 1984).

Huey, J. Executives assess Europe's technology decline. *Wall Street Journal* (Feb. 1, 1984).

Irving, R. I. Technology transfer: That's what it's all about. *Iron Age,* 53–62 (Nov. 7, 1983).

Jackall, R. Moral mazes: Bureaucracy and managerial work. *Harvard Business Review,* 118–130 (Sept.–Oct. 1983).

Jewkes, G., Thompson, P., and Dalton, G. How to stifle a technical organization

in ten easy steps. *Research Management,* 12–16 (Jan. 1979).
Juneau, L. M. Automation in the work place: Who's minding the shop? *The M.I.T. Report,* **12**(11) (Dec. 1984).
Kamien, Morton I., and Schwartz, Nancy L. *Market Structure and Innovation,* Cambridge, Mass: Cambridge University Press, 1982.
Kanter, Rose M. *The Change Masters,* New York: Simon and Schuster, 1983.
Karger, D. W., and Murdick, R. G. *Managing Engineering and Research,* New York: Industrial Press, Inc., 1980.
Kerzner, Harold, *Project Management: A Systems Approach to Planning, Scheduling, and Controlling,* New York: Van Nostrand Reinhold, 1979.
Kidder, Tracy, *The Soul of a New Machine,* Boston: Little, Brown, 1981.
Krawiec, F. Evaluating and selecting research projects by scoring. *Research Management,* 21–25 (Mar.–Apr. 1984).
Lal, Deepak. *Methods of Project Analysis: A Review,* Baltimore: Johns Hopkins University Press, 1974.
Lardner, J. Why U.S. manufacturers miss out on new technology. *Industry Week,* 13 (Aug. 23, 1982).
Lawless, Edward W. *Technology and Social Shock,* New Brunswick, N.J.: Rutgers University Press, 1977.
Leslie, S. W. When industrial research was young. *Research Management,* 11-19 (Sep.1982).
Lessem, R. Innovation and renewal. *Journal of General Management,* **9**(4):17–35 (Summer 1984).
Maciariello, Joseph A. *Program Management Control Systems,* New York: Wiley, 1978.
Malpas, R. The plant after next. *Harvard Business Review,* 122–130 (July–Aug. 1983).
Mansfield, Edwin. *The Production and Application of New Industrial Technology,* New York: Norton, 1977.
Martin, Charles C. *Project Management—How to Make It Work,* New York: American Management Association, 1976.
Martin, J. W. Managing small R & D projects—a learning model approach. *Research Management,* 15–21 (May 1980).
Mayo, J. Process design as important as product design. *Wall Street Journal* (Oct. 29, 1984).
McGinnis, M. A., and Ackelsberg, M. R. Effective innovation management: Missing link in strategic planning? *Journal of Business Strategy,* **4**(1):59–66 (Summer 1983).
Miller, J. A. From lab to shop floor—managing advanced manufacturing technology. *Mechanical Engineering,* **105**(11):62–68 (1983).
Miller, J. A. Manufacture of an electron beam welded turbine engine compressor rotor. *Welding Journal,* **56**(5):24–29 (May 1977).
Miller, J. A., and Chevalier, J. Development of an automated laser beam welding facility for high volume production. *Welding Journal,* **62**(7):49–54 (July 1983).
Miller, J. A., and O'Connor, J. J. Inertia and electron beam welded turbine engine power shaft. *Welding Journal,* **59**(5):17–22 (May 1980).
Moder, Joseph J., Phillips, Cecil R., and Davis, Edward W. *Project Management*

with CPM, PERT, and Precedence Diagramming (3rd ed.), New York: Van Nostrand Reinhold, 1983.

O'Boyle, T. F. U.S. basic industries are hindered by loss of scientific talent. *Wall Street Journal* (July 27, 1984).

Parker, J. E. S. *The Economics of Innovation,* London: Longman Group, Ltd., 1974.

Tuchman, Barbara W. *The March to Folly, from Troy to Vietnam.* New York: Knopf, 1984.

Walker, Charles R. *Modern Technology and Civilization,* New York: McGraw-Hill, 1962.

Willsmore, A. W. *Accounting for Management Control,* London: Pitman Publishing, 1971.

Winslow, R. Quality quandary—regulators investigate harrassing of inspectors at new nuclear plants. *Wall Street Journal* (Nov. 7, 1984).

Wolff, M. F. Bridging the R & D interface with manufacturing. *Research Management,* 9–11 (Jan.–Feb. 1985).

Wolff, M. F. Picking the right technology should be the first priority. *Research Management,* 7–8 (July 1981).

Wolff, M. F. Rules of thumb for project management, *Research Management,* 11–13 (July–Aug. 1984).

Woolf, Henry B. (Ed.). *Webster's New Collegiate Dictionary,* Springfield, Mass: Merriam, 1977.

Pavone, V.J. Methods for economic justification of an arc welding robot installation. *Welding Journal,* **62**(11):40–46 (Nov. 1983).

Peters, Thomas J., and Waterman, Robert H. *In Search of Excellence,* New York: Warner Books, 1982.

Petro, F. A. A new way out of the productivity and innovation morass? *Journal of Business Strategy,* **4**(2):42–48 (Fall 1983).

Podolsky, D. M. Vendor/user relationship plays key role in implementation. *Robotics World,* 6–8 (Oct. 1984).

Raisbeck, G. Systems development: Technology-push, user-pull, or producer-motivated. *Interface,* **12**(4) (Aug. 1982).

Raleigh, T. J. Adapting to a union-free environment. *Wall Street Journal* (Oct. 22, 1984).

Ranftl, Robert M. *R & D Productivity* (2nd ed.), Study Report, Hughes Aircraft Co., 1978.

Rosegger, Gerhard. *The Economics of Production and Innovation,* New York: Pergamon Press, 1980.

Rosenberg, Nathan (Ed.). *The Economics of Technological Change,* Bungay, England: Penguin Books, 1971.

Roussel, P. A. Cutting down guesswork in R & D. *Harvard Business Review,* 154–160 (Sept.–Oct. 1983).

Rowe, William D. *An Anatomy of Risk*, New York: Wiley, 1977.

Rudd, D. F., and Watson, C. C. *Strategy in Process Engineering,* New York: Wiley, 1968.

Scott, M., MacArthur, J. D., and Newberg, D. *Project Appraisal in Practice,* London: Heinemann Educational Books, 1976.

Seelig, W. D., and Rubin, I. M. *The Effects of PERT in R & D Organizations,* published as a working paper of the Research Program on the Management in Science and Technology, No. 230-66. Alfred P. Sloan School of Industrial Management, Massachusetts Institute of Technology, Cambridge, Mass., Dec. 1966. In Moder, Joseph T., Phillips, Cecil R., and Davis, Edward W. (Eds.), *Project Management with CPM, PERT, and Precedence Diagramming,* New York: Van Nostrand Reinhold, 1983.

Smith, J., et al. Lessons from 10 case studies in innovation—I. *Research Management,* 23–27 (Sept.–Oct. 1984).

Steele, L. Managing misconceptions about technology. *Harvard Business Review,* 133–140 (Nov.–Dec. 1983).

Steinberg, B. The military boost to industry. *Fortune Magazine,* 42–46 (April 30, 1984).

Stewart, R. *The Reality of Organizations,* Garden City, N.Y.: Doubleday, 1972.

Stoner, J. A. F. *Management,* Englewood Cliffs, N.J.: Prentice-Hall, 1982.

Strunk, William, Jr., and White, E. B. *The Elements of Style* (2nd ed.), New York: Macmillan, 1972.

INDEX